Basic Stochastic Processes

Series Editor
Jacques Janssen

Basic Stochastic Processes

Pierre Devolder
Jacques Janssen
Raimondo Manca

iSTE

WILEY

First published 2015 in Great Britain and the United States by ISTE Ltd and John Wiley & Sons, Inc.

ISTE Ltd
27-37 St George's Road
London SW19 4EU
UK

www.iste.co.uk

John Wiley & Sons, Inc.
111 River Street
Hoboken, NJ 07030
USA

www.wiley.com

Library of Congress Control Number: 2015942727

British Library Cataloguing-in-Publication Data
A CIP record for this book is available from the British Library
ISBN 978-1-84821-882-6

Contents

Introduction

This book will present basic stochastic processes for building models in insurance, especially in life and non-life insurance as well as credit risk for insurance companies. Of course, stochastic methods are quite numerous; so we have deliberately chosen to consider to use those induced by two big families of stochastic processes: stochastic calculus including Lévy processes and Markov and semi-Markov models. From the financial point of view, essential concepts such as the Black and Scholes model, VaR indicators, actuarial evaluation, market values and fair pricing play a key role, and they will be presented in this volume.

This book is organized into seven chapters. Chapter 1 presents the essential probability tools for the understanding of stochastic models in insurance. The next three chapters are, respectively, devoted to renewal processes (Chapter 2), Markov chains (Chapter 3) and semi-Markov processes both homogeneous and non-time homogeneous (Chapter 4) in time. This fact is important as new non-homogeneous time models are now becoming more and more used to build realistic models for insurance problems.

Chapter 5 gives the bases of stochastic calculus including stochastic differential equations, diffusion processes and changes of probability measures, therefore giving results that will be used in Chapter 6 devoted to Lévy processes. Chapter 6 is devoted to Lévy processes. This chapter also presents an alternative to basic stochastic models using Brownian motion as Lévy processes keep the properties of independent and stationary increments but without the normality assumption.

Finally, Chapter 7 presents a summary of Solvency II rules, actuarial evaluation, using stochastic instantaneous interest rate models, and VaR methodology in risk management.

Our main audience is formed by actuaries and particularly those specialized in entreprise risk management, insurance risk managers, Master's degree students in mathematics or economics, and people involved in Solvency II for insurance companies and in Basel II and III for banks. Let us finally add that this book can also be used as a standard reference for the basic information in stochastic processes for students in actuarial science.

1

Basic Probabilistic Tools for Stochastic Modeling

In this chapter, the readers will find a brief summary of the basic probability tools intensively used in this book. A more detailed version including proofs can be found in [JAN 06].

1.1. Probability space and random variables

Given a sample space Ω, the set of all possible events will be denoted by $\Im$, which is assumed to have the structure of a σ*-field* or a σ*-algebra*. P will represent a probability measure.

DEFINITION 1.1.– *A random variable (r.v.) with values in a topological space* (E,ψ) *is an application X from* Ω *to E such that:*

$$\forall B \in \psi : X^{-1}(B) \in \Im, \qquad [1.1]$$

where $X^{-1}(B)$ is called the inverse image of the set B defined by:

$$X^{-1}(B) = \{\omega : X(\omega) \in B\}, X^{-1}(B) \in \Im . \qquad [1.2]$$

Particular cases:

a) If $(E,\psi) = (\mathbb{R}, \beta)$, X is called a *real random variable.*

b) If $(E,\psi)=(\overline{\mathbb{R}},\overline{\beta})$, where $\overline{\mathbb{R}}$ is the *extended real line* defined by $\mathbb{R}\cup\{+\infty\}\cup\{-\infty\}$ and $\overline{\beta}$ is the *extended Borel σ -field* of $\overline{\mathbb{R}}$, that is the minimal σ -field containing all the elements of β and the extended intervals:

$$\begin{aligned}&[-\infty,a),(-\infty,a],[-\infty,a],(-\infty,a),\\&[a,+\infty),(a,+\infty],[a,+\infty],(a,+\infty),\ a\in\mathbb{R},\end{aligned} \qquad [1.3]$$

X is called a *real extended value* random variable.

c) If $E=\mathbb{R}^n (n>1)$ with the product σ -field $\beta^{(n)}$ of β, X is called an *n-dimensional real random variable.*

d) If $E=\overline{\mathbb{R}}^{(n)}$ (n>1) with the product σ -field $\overline{\beta}^{(n)}$ of $\overline{\beta}$, X is called a real *extended n-dimensional real random variable.*

A random variable X is called *discrete* or *continuous* accordingly as X takes at most a denumerable or a non-denumerable infinite set of values.

DEFINITION 1.2.– *The distribution function of the r.v. X, represented by F_X, is the function from $\mathbb{R}\to[0,1]$ defined by:*

$$F_X(x)=P(\{\omega: X(\omega)\le x\}). \qquad [1.4]$$

Briefly, we write:

$$F_X(x)=P(X\le x). \qquad [1.5]$$

This last definition can be extended to the multi-dimensional case with a r.v. X being an n-dimensional real vector: $X=(X_1,\ldots,X_n)$, a measurable application from $(\Omega,\Im,P)$ to $(\mathbb{R}^n,\beta^n)$.

DEFINITION 1.3.– *The distribution function of the r.v. $X=(X_1,\ldots,X_n)$, represented by F_X, is the function from $\mathbb{R}^n$ to $[0,1]$ defined by:*

$$F_X(x_1,\ldots,x_n)=P(\{\omega: X_1(\omega)\le x_1,\ldots,X_n(\omega)\le x_n\}). \qquad [1.6]$$

Briefly, we write:

$$F_X(x_1,\dots,x_n) = P(X_1 \le x_1,\dots,X_n \le x_n). \quad [1.7]$$

Each component X_i ($i = 1,\dots,n$) is itself a one-dimensional real r.v. whose d.f., called the *marginal d.f.*, is given by:

$$F_{X_i}(x_i) = F_X(+\infty,\dots,+\infty,x_i,+\infty,\dots,+\infty). \quad [1.8]$$

The concept of random variable is *stable* under a lot of mathematical operations; so any Borel function of a r.v. X is also a r.v.

Moreover, if X and Y are two r.v., so are:

$$\inf\{X,Y\},\sup\{X,Y\},X+Y,X-Y,X\cdot Y,\frac{X}{Y}, \quad [1.9]$$

provided, in the last case, that Y does not vanish.

Concerning the convergence properties, we must mention the property that, if $(X_n, n \ge 1)$ is a *convergent* sequence of r.v. – that is, for all $\omega \in \Omega$, the sequence $(X_n(\omega))$ converges to $X(\omega)$ – then the limit X is also a r.v. on Ω. This convergence, which may be called the *sure convergence,* can be weakened to give the concept of *almost sure* (a.s.) *convergence* of the given sequence.

DEFINITION 1.4.– *The sequence* $(X_n(\omega))$ *converges a.s. to* $X(\omega)$ *if:*

$$P(\{\omega: \lim X_n(\omega) = X(\omega)\}) = 1 \quad [1.10]$$

This last notion means that the possible set where the given sequence does not converge is a *null* set, that is, a set N belonging to $\Im$ such that:

$$P(N) = 0. \quad [1.11]$$

In general, let us remark that, given a null set, it is not true that every subset of it belongs to $\Im$ but of course if it belongs to $\Im$, it is clearly a null set. To avoid unnecessary complications, we will assume from here onward that any considered probability space is *complete,* i.e. all the subsets of a null set also belong to $\Im$ and thus their probability is zero.

1.2. Expectation and independence

Using the concept of integral, it is possible to define the *expectation* of a random variable X represented by:

$$E(X) = \int_{\Omega} XdP \left(= \int XdP\right), \quad [1.12]$$

provided that this integral exists. The computation of the integral:

$$\int_{\Omega} XdP \left(= \int XdP \right) \quad [1.13]$$

can be done using the induced measure μ on $(\mathbb{R}, \beta)$, defined by [1.4] and then using the distribution function F of X.

Indeed, we can write:

$$E(X) \left(= \int_{\Omega} XdP \right), \quad [1.14]$$

and if F_X is the d.f. of X, it can be shown that:

$$E(X) = \int_{R} xdF_X(x). \quad [1.15]$$

The last integral is a Lebesgue–Stieltjes integral.

Moreover, if F_X is absolutely continuous with f_X as density, we obtain:

$$E(X) = \int_{-\infty}^{+\infty} xf_x(x)dx. \quad [1.16]$$

If g is a Borel function, then we also have (see, e.g. [CHU 00] and [LOÈ 63]):

$$E(g(X)) = \int_{-\infty}^{+\infty} g(x)dF_X \quad [1.17]$$

and with a density for X:

$$E(g(X)) = \int_{-\infty}^{+\infty} g(x) f_X(x) dx . \quad [1.18]$$

It is clear that the expectation is a linear operator on integrable functions.

DEFINITION 1.5.– *Let a be a real number and r be a positive real number, then the expectation:*

$$E\left(|X-a|^r\right) \quad [1.19]$$

is called the absolute moment of X, of order r, centered on a.

The moments are said to be centered moments of order r if $a=E(X)$. In particular, for $r = 2$, we get the *variance* of X represented by σ^2 (var(X)):

$$\sigma^2 = E\left(|X-m|^2\right). \quad [1.20]$$

REMARK 1.1.– From the linearity of the expectation, it is easy to prove that:

$$\sigma^2 = E(X^2) - (E(X))^2, \quad [1.21]$$

and so:

$$\sigma^2 \le E(X^2), \quad [1.22]$$

and, more generally, it can be proved that the variance is the smallest moment of order 2, whatever the number a is.

The set of all real r.v. such that the moment of order r exists is represented by L^r.

The last fundamental concept that we will now introduce in this section is *stochastic independence*, or more simply *independence*.

DEFINITION 1.6.– *The events* $A_1,\ldots,A_n,(n>1)$ *are stochastically independent or independent iff:*

$$\forall m = 2,\ldots,n, \forall n_k = 1,\ldots,n : n_1 \neq n_2 \neq \cdots \neq n_k : P\left(\bigcap_{k=1}^{m} A_{n_k}\right) = \prod_{k=1}^{m} P(A_{n_k}). \quad [1.23]$$

For $n = 2$, relation [1.23] reduces to:

$$P(A_1 \cap A_2) = P(A_1)P(A_2). \qquad [1.24]$$

Let us remark that piecewise independence of the events $A_1,..., A_n,(n>1)$ does not necessarily imply the independence of these sets and, thus, not the stochastic independence of these n events.

From relation [1.23], we find that:

$$P(X_1 \le x_1,...,X_n \le x_n) = P(X_1 \le x_1)\cdots P(X_n \le x_n), \forall (x_1,...,x_n) \in \mathbb{R}^n. \qquad [1.25]$$

If the functions $F_X, F_{X_1},..., F_{X_n}$ are the distribution functions of the r.v. $X = (X_1,...,X_n), X_1,...,X_n$, we can write the preceding relation as follows:

$$F_X(x_1,....,x_n) = F_{X_1}(x_1)\cdot\cdots\cdot F_{X_n}(x_n), \forall (x_1,...,x_n) \in \mathbb{R}^n. \qquad [1.26]$$

It can be shown that this last condition is also sufficient for the independence of $X = (X_1,...,X_n), X_1,...,X_n$. If these d.f. have densities $f_X, f_{X_1},..., f_{X_n}$, relation [1.24] is equivalent to:

$$f_X(x_1,...,x_n) = f_{X_1}(x_1)\cdots f_{X_n}(x_n), \forall (x_1,...,x_n) \in \mathbb{R}^n. \qquad [1.27]$$

In case of the integrability of the n real r.v $X_1,X_2,...,X_n$, a direct consequence of relation [1.26] is that we have a very important property for the expectation of the product of n independent r.v.:

$$E\left(\prod_{k=1}^{n} X_k\right) = \prod_{k=1}^{n} E(X_k). \qquad [1.28]$$

The notion of independence gives the possibility of proving the result called the *strong law of* large *numbers*, which states that if $(X_n, n \ge 1)$ is a sequence of integrable independent and identically distributed r.v., then:

$$\frac{1}{n}\sum_{k=1}^{n} X_k \xrightarrow{a.s.} E(X). \qquad [1.29]$$

The next section will present the most useful distribution functions for stochastic modeling.

DEFINITION 1.7 (SKEWNESS AND KURTOSIS COEFFICIENTS).–

a) The *skewness coefficient of Fisher* is defined as follows:

$$\gamma_1 = \frac{E\left[(X - E(X))^3\right]}{\sigma^3}$$

From the odd value of this exponent, it follows that:

– $\gamma_1 > 0$ gives a *left dissymmetry* giving a maximum of the density function situated to the left and a distribution with a right heavy queue, $\gamma_1 = 0$ gives *symmetric* distribution with respect to the mean;

– $\gamma_1 < 0$ gives a *right dissymmetry* giving a maximum of the density function situated to the right and a distribution with a left heavy queue.

b) The *kurtosis coefficient* also due to Fisher is defined as follows:

$$\gamma_2 = \frac{E\left[(X - E(X))^4\right]}{\sigma^4}$$

Its interpretation refers to the normal distribution for which its value is 3. Also some authors refer to the *excess of kurtosis* given by γ_1-3 of course null in the normal case.

For $\gamma_2 < 3$, distributions are called *leptokurtic*, being more plated around the mean than in the normal case and with heavy queues.

For $\gamma_2 > 3$, distributions are less plated around the mean than in the normal case and with heavy queues.

1.3. Main distribution probabilities

In this section, we will restrict ourselves to presenting the principal distribution probabilities related to real random variables.

1.3.1. *Binomial distribution*

Let X be a discrete random variable, whose distribution $(p_i, i = 0,...,n)$ with:

$$p_i = P(X = i),\ i = 1,...,n \qquad [1.30]$$

is called a *binomial* distribution with parameters (n,p) if:

$$p_i = \binom{n}{i} p^i q^{n-i}, \; i = 0,\dots,n, \qquad [1.31]$$

a result from which we get:

$$E(X) = np, \mathrm{var}(X) = npq. \qquad [1.32]$$

The characteristic function and the generating function, when the latter exists, of X, respectively, defined by:

$$\begin{aligned} \varphi_X(t) &= E(e^{itX}), \\ g_X(t) &= E(e^{tX}) \end{aligned} \qquad [1.33]$$

are given by:

$$\begin{aligned} \varphi_X(t) &= (pe^{it} + q)^n, \\ g_X(t) &= (pe^{t} + q)^n. \end{aligned} \qquad [1.34]$$

1.3.2. *Negative exponential distribution*

This is defined by:

$$\begin{aligned} &\Pr[X \le x] = \begin{cases} 0 & x \le 0 \\ 1 - e^{-\frac{1}{\beta}x} & x \ge 0 \end{cases} \\ &E[X] = \beta, \\ &\text{variance}[X] = \beta^2, \\ &\gamma_1 = 2, \gamma_2 = 9. \end{aligned} \qquad [1.35]$$

1.3.3. *Normal (or Laplace–Gauss) distribution*

The real r.v. X has a normal (or Laplace–Gauss) distribution of parameters $(\mu, \sigma^2), \mu \in \mathbb{R}, \sigma^2 > 0$, if its density function is given by:

$$f_X(x)=\frac{1}{\sqrt{2\pi}}e^{-\frac{(x-\mu)^2}{2\sigma^2}}, x\in\mathbb{R}. \quad [1.36]$$

From now on, we will use the notation $X \prec N(\mu,\sigma^2)$.

The main parameters of this distribution are:

$$E(X)=\mu,\ \text{var}(X)=\sigma^2,$$
$$\varphi_X(t)=\exp\left(i\mu t-\frac{\sigma^2t^2}{2}\right),\ g_X(t)=\exp\left(\mu t+\frac{\sigma^2t^2}{2}\right). \quad [1.37]$$

If $\mu=0,\ \sigma^2=1$, the distribution of X is called a *reduced* or *standard normal distribution.* In fact, if X has a normal distribution $(\mu,\sigma^2),\mu\in R,\sigma^2>0$, then the so-called reduced r.v. Y defined by:

$$Y=\frac{X-\mu}{\sigma} \quad [1.38]$$

has a standard normal distribution, thus from [1.36] with mean 0 and variance 1.

Let $\exists k>0: \lim_{x\to\infty}\frac{1-F(x)}{1-F(ux)}=u^k$, for all $u>0$ be the distribution function of the standard normal distribution; it is possible to express the distribution function of any normal r.v. X of parameters $(\mu,\sigma^2),\mu\in\mathbb{R},\sigma^2>0$, as follows:

$$F_X(x)=P(X\le x)=P\left(\frac{X-\mu}{\sigma}\le\frac{x-\mu}{\sigma}\right)=\Phi\left(\frac{x-\mu}{\sigma}\right). \quad [1.39]$$

In addition, from the numerical point of view, it is sufficient to know the numerical values for the standard distribution.

From relation [1.38], we also deduce that:

$$f_X(x)=\frac{1}{\sigma}\Phi'\left(\frac{x-\mu}{\sigma}\right), \quad [1.40]$$

where, of course, from [1.35]:

$$\Phi'(x) = \frac{1}{\sqrt{2\pi}} e^{-\frac{x^2}{2}}. \quad [1.41]$$

From the definition of Φ, we have:

$$\Phi(x) = \frac{1}{\sqrt{2\pi}} \int_{-\infty}^{x} e^{-\frac{y^2}{2}} dy, \ x \in \mathbb{R} \quad [1.42]$$

and so:

$$\Phi(-x) = 1 - \Phi(x), x > 0, \quad [1.43]$$

and consequently, for X normally distributed with parameters (0,1), we obtain:

$$P(|X| \leq x) = \Phi(x) - \Phi(-x) = 2\Phi(x) - 1, \ x > 0. \quad [1.44]$$

In particular, let us mention the following numerical results:

$$\begin{aligned}
&P\left(|X - m| \leq \frac{2}{3}\sigma\right) = 0.4972(\approx 50\%), \\
&P(|X - m| \leq \sigma) = 0.6826(\approx 68\%), \\
&P(|X - m| \leq 2\sigma) = 0.9544(\approx 95\%), \\
&P(|X - m| \leq 3\sigma) = 0.9974(\approx 99\%).
\end{aligned} \quad [1.45]$$

The normal distribution is one of the most often used distributions by virtue of the *central limit theorem*, which states that if $(X_n, n \geq 1)$ is a sequence of independent and identically distributed (i.i.d.) r.v. with mean m and variance σ^2, then the sequence of r.v. is defined by:

$$\frac{S_n - nm}{\sigma\sqrt{n}} \quad [1.46]$$

with:

$$S_n = X_1 + \cdots + X_n, \quad n > 0 \quad [1.47]$$

converging in law to a standard normal distribution.

This means that the sequence of the distribution functions of the variables defined by [3.20] converges to Φ.

This theorem was used by the Nobel Prize winner H. Markowitz [MAR 159] to justify that the return of a diversified portfolio of assets has a normal distribution. As a particular case of the central limit theorem, let us mention the *de Moivre's theorem* starting with:

$$X_n = \begin{cases} 1, \text{ with prob. } p, \\ 0, \text{ with prob. } 1-p, \end{cases} \quad [1.48]$$

so that, for each n, the r.v. defined by relation [1.47] has a binomial distribution with parameters (n,p).

Now by applying the central limit theorem, we get the following result:

$$\frac{S_n - np}{\sqrt{np(1-p)}} \xrightarrow[n\to+\infty]{law} N(0,1), \quad [1.49]$$

which is called de Moivre's result.

1.3.4. *Poisson distribution*

This is a discrete distribution with the following characteristics:

$$\begin{aligned} &P(\xi = n) = e^{-\lambda}\frac{\lambda^n}{n!}, n = 0,1,... \\ &m = \sigma^2 = \lambda, \\ &\gamma_1 = \frac{1}{\sqrt{\lambda}}, \gamma_2 = \frac{1}{\lambda} + 3. \end{aligned} \quad [1.50]$$

This is one of the most important distributions for all applications. For example, if we consider an insurance company looking at the total number of claims in one year, this variable may often be considered to be a Poisson variable.

1.3.5. *Lognormal distribution*

This is a continuous distribution on the positive half line with the following characteristics:

$$\Pr[\ln X \le x] = \frac{1}{\sqrt{2\pi}\sigma} e^{-\frac{(x-\mu)^2}{2\sigma^2}}$$

$$E[X] = e^{\mu+\frac{\sigma^2}{2}},$$

$$\text{variance}[X] = e^{2\mu+\sigma^2}\left(e^{\sigma^2}-1\right), \tag{1.51}$$

$$\gamma_1 = (e^{\sigma^2}+2)\sqrt{(e^{\sigma^2}-1)}$$

$$\gamma_2 = \omega^4 + 2\omega^3 + 3\omega^2 - 3, \omega = e^{\sigma^2}.$$

Let us say that the lognormal distribution has no generating function and that the characteristic function has no explicit form. When $\sigma < 0.3$, some authors recommend a normal approximation with parameters (μ, σ^2).

The normal distribution is *stable* under the addition of independent random variables; this property means that the sum of n independent normal r.v. is still normal. This is no longer the case with the lognormal distribution which is stable under *multiplication*, which means that for two independent lognormal r.v. X_1, X_2, we have

$$X_i \prec LN(\mu_i, \sigma_i), i = 1,2 \Rightarrow X_1 \times X_2 \prec LN\left(\mu_1 + \mu_2, \sqrt{\sigma_1^2 + \sigma_2^2}\right). \tag{1.52}$$

1.3.6. *Gamma distribution*

This is a continuous distribution on the positive half line having the following characteristics:

$$\Pr[x < X \le x + \Delta x] = \frac{\theta^\nu}{(\nu-1)!} x^{\nu-1} e^{-\theta x} \Delta x$$

$$E[X] = \frac{\nu}{\theta},$$

$$\text{variance}[X] = \frac{\nu}{\theta^2}, \tag{1.53}$$

$$\gamma_1 = 2/\sqrt{\nu},$$

$$\gamma_2 = \frac{6}{\nu} + 3.$$

For the gamma law with parameters (ν, θ) denoted $\gamma(\nu, \theta)$, an additivity property exists:

$$\gamma(\nu, \theta) + \gamma(\nu', \theta) \succ \gamma(\nu + \nu', \theta).$$

1.3.7. *Pareto distribution*

The non-negative r.v. X has a *Pareto distribution* if its distribution function is given by:

$$F_X(x) = 1 - \left(\frac{k}{x}\right)^{\alpha}, \quad x > k;\ k > 0,\ \alpha > 0. \qquad [1.54]$$

Its support is $(k, +\infty)$.

The corresponding density function is given by:

$$f_X(x) = \frac{\alpha k^{\alpha}}{x^{\alpha+1}}, x \geq k. \qquad [1.55]$$

The Pareto distribution has centered moments of order r provided that $r < \alpha$; and in this case:

$$E\left[X^r\right] = \frac{\alpha k^r}{\alpha - r}, \ r < \alpha$$

and so:

$$E[X] = \frac{\alpha k}{\alpha - 1}, \alpha > 1,$$

$$\operatorname{var} X = \frac{\alpha k^2}{(\alpha-1)^2(\alpha-2)}, \alpha > 2.$$

$$\gamma_1 = \frac{\alpha+1}{\alpha-3}\sqrt{1-\frac{2}{\alpha}}, \alpha > 3; \gamma_2 = \frac{3(\alpha-2)(3\alpha^2+\alpha+2)}{\alpha(\alpha-3)(\alpha-4)}, \alpha > 4.$$

These values explain why this distribution is considered to be *dangerous* as for some values of the parameters it is not excluded to observe large values of X in a random experiment.

For $\alpha < 1$, the mean is infinite, and for $1 < \alpha < 2$, although the mean is finite, the variance is not.

The problem of this distribution also comes from the fact that the function 1-$F(x)$ decreases in a polynomial way for large x (distribution with heavy queue) and no longer exponentially like the other presented distributions, except, of course, for the Cauchy distribution.

In non-life insurance, it is used for modeling large claims and catastrophic events.

REMARK 1.2.– We also have:

$$\ln(1 - F_X(x)) = \ln\left(\frac{k}{x}\right)^{\alpha}, \; x > k;\; k > 0,\; \alpha > 0,$$

ou

$$\ln(1 - F_X(x)) = \alpha(\ln k - \ln x).$$

REMARK 1.3.– If we compare the form:

$$F_X(x) = 1 - \left(\frac{k}{x}\right)^{\alpha}, \quad x > k;\; k > 0,\; \alpha > 0. \qquad [1.56]$$

with:

$$F_{X'}(x) = \begin{cases} 1 - \left(\dfrac{\theta}{x+\theta}\right)^{\beta}, & x \geq 0, \\ 0, & x < 0, \end{cases} \qquad [1.57]$$

we have another form of the Pareto distribution with as support all the positive half line. This is possible with the change of variable:

Y=X+k

Of course, the variances remain the same but not the means.

$$m_X = \frac{\alpha k}{\alpha - 1} (\alpha > 1),\; m_{X'} = \frac{\theta}{\beta - 1} (\beta > 1) \qquad [1.58]$$

$$\sigma_X^2 = \frac{\alpha k^2}{(\alpha-1)^2(\alpha-2)}(\alpha > 2), \qquad \sigma_{X'}^2 = \frac{\theta^2 \beta}{(\beta-1)^2(\beta-2)}(\beta > 2) = \frac{m_{X'}^2 \beta}{\beta-2}$$

Here are two graphs of the distribution function showing the impact of the dangerous parameters.

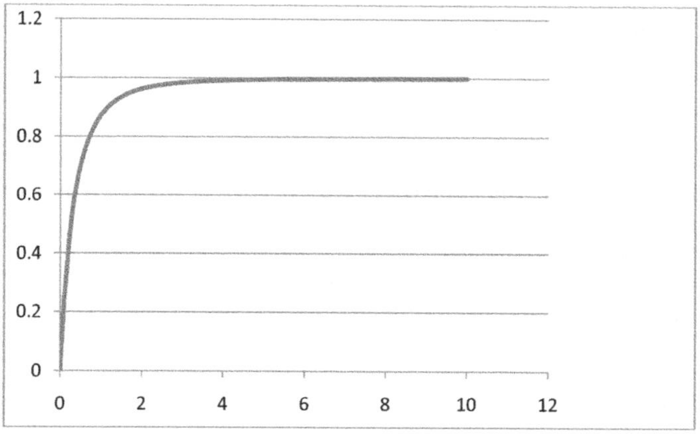

Figure 1.1. *Pareto distribution function with Θ=1,β=1*

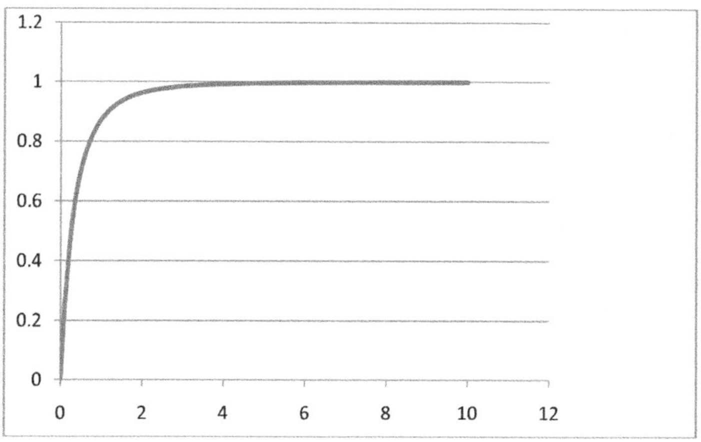

Figure 1.2. *Pareto distribution with β=3,θ=1*

REMARK 1.4.– As we have:

$$\ln(1-F_X(x)) = \ln\left(\frac{k}{x}\right)^{\alpha}, \; x>k;\; k>0,\; \alpha>0,$$

or [1.59]

$$\ln(1-F_X(x)) = \alpha(\ln k - \ln x).$$

The proportion of claims larger than x is a linear function of x in a double logarithmic scale with α as slope.

1.3.8. *Uniform distribution*

Its support is [a,b] on which the density is constant with the value 1/(b-a).

For basic parameters, we have:

$$m = \frac{b-a}{2}, \sigma^2 = \frac{(b-a)^2}{12}$$
$$\gamma_1 = 0, \gamma_2 = 1,8.$$

1.3.9. *Gumbel distribution*

This is related to a non-negative random variable with the following characteristics:

$$\begin{aligned}
&F(x) = e^{-e^{-x}},\\
&f(x) = e^{(-x-e^{-x})},\\
&E(X) = 0,57722...,\\
&\operatorname{var}(X) = \frac{\pi^2}{6},\\
&\gamma_1 = 1,29857, \gamma_2 = 5,4.
\end{aligned} \qquad [1.60]$$

1.3.10. *Weibull distribution*

This is related to a non-negative random variable with the following characteristics:

$$
\begin{aligned}
&F(x)=1-e^{-(\frac{x}{\alpha})^{\beta}},\alpha,\beta>0.\\
&f(x)=\alpha\beta^{-\alpha}x^{\alpha-1}e^{-(\frac{x}{\alpha})^{\beta}},\\
&E(X)=\frac{\beta}{\alpha}\Gamma(\frac{1}{\alpha}),\\
&\operatorname{var}(X)=\frac{\beta^2}{\alpha}\left[2\Gamma(\frac{2}{\alpha})-\frac{1}{\alpha}(\Gamma(\frac{1}{\alpha})^2\right].
\end{aligned}
\quad [1.61]
$$

1.3.11. *Multi-dimensional normal distribution*

Let us consider an n-dimensional real r.v. X represented as a column vector of its n components $X=(X_1,...,X_n)'$. Its d.f. is given by:

$$F_X(x_1,...,x_n)=P(X_1\le x_1,...,X_n\le x_n). \quad [1.62]$$

If the density function of X exists, the relations between the d.f. and the density function are:

$$
\begin{aligned}
&f_X(x_1,...,x_n)=\frac{\partial^n F_X}{\partial x_1...\partial x_n}(x_1,...,x_n),\\
&F_X(x_1,...,x_n)=\int_{-\infty}^{x_1}...\int_{-\infty}^{x_n}f_X(\xi_1,...,\xi_n)d\xi_1,...,d\xi_n.
\end{aligned}
\quad [1.63]
$$

For the principal parameters, we will use the following notation:

$$
\begin{aligned}
&E(X_k)=\mu_k,k=1,...,n,\\
&E((X_k-\mu_k)(X_l-\mu_l))=\sigma_{kl},k,l=1,...,n,\\
&E((X_k-\mu_k))^2=\sigma_k^2,k=1,...,n,\\
&\rho_{kl}=\frac{E((X_k-\mu_k)(X_l-\mu_l))}{\sqrt{E((X_k-\mu_k)^2)E((X_k-\mu_k)^2)}}\left(=\frac{\sigma_{kl}}{\sigma_k\sigma_l}\right),k,l=1,...,n.
\end{aligned}
\quad [1.64]
$$

The parameters σ_{kl} are called the *covariances* between the r.v. X_k and X_l and the parameters ρ_{kl} are called the *correlation coefficients* between the r.v. X_k and X_l.

It is well known that the correlation coefficient ρ_{kl} measures a certain linear dependence between the two r.v. X_k and X_l. More precisely, if it is equal to 0, then there is no such dependence and the two variables are called *uncorrelated*; for the values +1 and −1, this dependence is certain.

With matrix notation, the following σ matrix:

$$\boldsymbol{\Sigma}_X = \left[\sigma_{ij}\right] \tag{1.65}$$

is called the *variance*–covariance matrix of X.

The characteristic function of X is defined as:

$$\varphi_X(t_1,\ldots,t_n) = E\left(e^{i(t_1X_1+..+t_nX_n)}\right)\left(= E\left(e^{i\mathbf{t}'X}\right)\right). \tag{1.66}$$

Let $\boldsymbol{\mu},\boldsymbol{\Sigma}$ be, respectively, an n-dimensional real vector and an $n\times n$ positive definite matrix. The n-dimensional real r.v. X has a *non-degenerated n-dimensional normal distribution* with parameters $\boldsymbol{\mu},\boldsymbol{\Sigma}$ if its density function is given by:

$$f_X(\mathbf{x}) = \frac{1}{(2\pi)^{\frac{n}{2}}\sqrt{\det\boldsymbol{\Sigma}}} e^{-\frac{1}{2}(\mathbf{x}-\boldsymbol{\mu})'\boldsymbol{\Sigma}^{-1}(\mathbf{x}-\boldsymbol{\mu})}, \mathbf{x}\in\mathbb{R}^n. \tag{1.67}$$

Then, it can be shown by integration that parameters $\boldsymbol{\mu},\boldsymbol{\Sigma}$ are indeed, respectively, the *mean vector* and the *variance–covariance matrix* of X.

As usual, we will use the notation $X \prec N_n(\boldsymbol{\mu},\boldsymbol{\Sigma})$.

The characteristic function of X is given by:

$$\varphi_X(\mathbf{t}) = e^{i\boldsymbol{\mu}'\mathbf{t}-\frac{1}{2}\mathbf{t}'\boldsymbol{\Sigma}\mathbf{t}}. \tag{1.68}$$

The main fundamental properties of the n-dimensional normal distribution are:

– every subset of k r.v. of the set $\{X_1,\ldots,X_n\}$ also has a k-dimensional distribution which is also normal;

– the multi-dimensional normal distribution is *stable* under linear transformations of X and for the addition, we have that if $X_k \prec N_n(\boldsymbol{\mu}_k,\boldsymbol{\Sigma}_k), k=1,\ldots,m$ and if these m random vectors are independent, then:

$$X_1+\cdots+X_m \prec N_n(\boldsymbol{\mu}_1+\cdots+\boldsymbol{\mu}_m,\boldsymbol{\Sigma}_1+\cdots+\boldsymbol{\Sigma}_m). \tag{1.69}$$

In the particular case of the *two-dimensional normal distribution*, we have:

$$\boldsymbol{\mu} = (\mu_1, \mu_2)', \boldsymbol{\Sigma} = \begin{bmatrix} \sigma_1^2 & \sigma_{12} \\ \sigma_{21} & \sigma_2^2 \end{bmatrix}, \rho = \frac{\sigma_{12}}{\sigma_1 \sigma_2},$$
$$\boldsymbol{\Sigma}^{-1} = \begin{bmatrix} \dfrac{1}{\sigma_1^2} & -\dfrac{\rho}{\varsigma_1 \sigma_2} \\ -\dfrac{\rho}{\varsigma_1 \sigma_2} & \dfrac{1}{\sigma_2^2} \end{bmatrix}, \det \boldsymbol{\Sigma} = \sigma_1 \sigma_2 \sqrt{1-\rho^2}. \quad [1.70]$$

From the first main fundamental properties of the n-dimensional normal distribution given above, we deduce that:

$$X_k \prec N_1(\mu_k, \sigma_k^2), k = 1, 2. \quad [1.71]$$

For the special degenerated case $|\rho| = 1$, it can be proved that:

$$\rho = 1: \frac{X_2 - \mu_2}{\sigma_2} = \frac{X_1 - \mu_1}{\sigma_1},$$
$$\rho = -1: \frac{X_2 - \mu_2}{\sigma_2} = -\frac{X_1 - \mu_1}{\sigma_1}, \quad [1.72]$$

relations meaning that, in this case, the entire probability mass in the plan lies on a straight line so that the two random variables X_1 and X_2 are perfectly dependent, i.e. relations [1.72] are true with probability one.

To conclude this section, let us recall the well-known property saying that two independent r.v. are uncorrelated but the converse is not true except for the normal distribution.

1.3.12. *Extreme value distribution*

In this section, we present basic results on the theory of extreme values [EMB 08], well adapted for the large claims designing risks which can take large values even with small probabilities but far from the mean value.

1.3.12.1. *Definition*

Let $X_1, \ldots, X_n$ be the independent realizations of the considered risk X and the risk of this sample can be measured by the largest claim value that is:

$$Z_n = \max\{X_1,...,X_n\}. \quad [1.73]$$

If F is the distribution function of the r.v. X, we have from the independence assumption:

$$P(Z_n \le z) = F^n(x). \quad [1.74]$$

As this probability tends to 0 with $n \to \infty$ and for all x, it is necessary to find asymptotic results giving a more precise view of what could happen for large n and that is, in fact, the aim of extreme theory.

1.3.12.2. *Asymptotic results*

Fisher [FIS 28] and Gnedenko [GNE 43] proved that if there exist two sequences of real numbers $(c_n),(d_n)$ with $X_1,...,X_n > 0$ for all n, such that the distribution of the following r.v.:

$$Y_n = \frac{\max\{X_1,...X_n\} - d_n}{c_n} \quad [1.75]$$

is not degenerated, then the limit distribution of Y_n must have one of the three following forms:

i) Gumbel's law: $\Lambda(x) = e^{-e^{-x}}, x \in \mathbb{R}$. [1.76]

ii) Fréchet's law: $\Phi_\beta(x) = \begin{cases} 0, x \le 0, \\ e^{-x^{-\beta}}, x > 0, \end{cases}, \beta > 0.$ [1.77]

iii) Weibull's law: $\psi_\beta(x) = \begin{cases} e^{-(-x)^\beta}, x < 0, \\ 1, \quad x \ge 0, \end{cases} \beta > 0.$ [1.78]

If we introduce the function

$$a(y) = \exp\left[-(1-\tau y)^{1/\tau}\right], \quad [1.79]$$

($\tau = -\dfrac{1}{\beta}$ for Fréchet, $\tau = \dfrac{1}{\beta}$ for Weinbull and $\tau = 0$ for Gumbel),

the preceding characterizations of the three attractions domains can be given as follows:

$$\text{Fréchet}: F(y) = \begin{cases} 0, y \le 1/\tau \\ a(y), y > 1/\tau \end{cases} \tau < 0,$$

$$\text{Gumbel}: F(y) = a(y) = e^{-e^{-y}}, \tau = 0, \quad [1.80]$$

$$\text{Weibull}: F(y) = \begin{cases} a(y), y < 1/\tau \\ 1, y \ge 1/\tau \end{cases} \tau > 0.$$

To verify this result, for example, for the Fréchet law, we introduce z and k defined as:

$$x = 1 - \tau y, \beta = -\frac{1}{\tau}$$

with $\tau < 0$ as k is positive and $y > \frac{1}{\tau}$ as $x > 0$. Moreover, as from [1.77], we obtain well:

$$\Phi_\alpha(x) = \begin{cases} 0, y \le \frac{1}{\tau}, \\ e^{-(1-\tau y)^{1/\tau}}, y > \frac{1}{\tau}. \end{cases} \quad [1.81]$$

For the Weibull case, we have to define x and β as:

$$x = \tau y - 1, \beta = \frac{1}{\tau}$$

And for the Gumbel case, we have:

$$\lim_{\tau \searrow 0} \exp\left[-(1-\tau y)^{1/\tau}\right] = \exp\left[-\lim_{n\to\infty}(1-\frac{y}{n})^n\right]$$
$$= e^{-e^{-y}}.$$

REMARK 1.5.–

i) By Taylor expansion, we have for the Fréchet law:

$$\Phi_\beta(x) \approx x^{-\beta}, x \to \infty \quad [1.82]$$

and so the tail of Φ_β decreases like a power law.

ii) We have the following equivalences:

$$H \prec \Phi_\beta \Leftrightarrow \ln X^\beta \prec \Delta \Leftrightarrow -X^{-1} \prec \Psi_\beta \quad [1.83]$$

The parameter $E(Y|\Im_1) = E(Y|B)$ can be seen as a dispersion parameter and d_n as a localization parameter tending toward the mode, i.e. the maximum of the density function of the limit distribution. Gnedenko [GNE 43] characterized the three classes of the distribution function F of the considered risk called *attraction domains*; so if a risk has its distribution function in one of these three domains, we know what the limit distribution for this risk is.

To characterize these three attraction domains, let us introduce the concept of slowly varying function L.

Such a function with support $(0,\infty)$ is *slowly varying type* iff:

$$\lim_{x\to\infty} \frac{L(tx)}{L(x)} = 1, \forall t > 0 \quad [1.84]$$

Moreover, if:

$$\lim_{x\to\infty} \frac{L(tx)}{L(x)} = t^\alpha, \forall t > 0 \quad [1.85]$$

L is called *regularly varying at index α.*

We can now give the following characterization of d the three attraction domains (see [EMB 08]):

i) The *Gumbel attraction* domain contains the distribution functions F so that:

$\exists x_0$:

$F(x_0) = 1$ et $F(x) < 1$ for $x < x_0$

and a $z < x_0$ such that :

ii) $$\bar{F}(x) = c(x)\exp\left\{-\int_0^x \frac{g(t)}{a(t)} dt\right\}, z < x < x_0 \quad [1.86]$$

where c and g are measurable functions such that

$\lim_{x\nearrow x_0} c(x) = c > 0$ and $\lim_{x\nearrow x_0} g(x) = 1 > 0$

and $a(x)$a positive absolutely function with Lebesgue

density a' such that $\lim_{x\nearrow x_0} a'(x) = 0$ as for example

$$\text{a(x)}=\int_x^{x_0} \frac{\bar{F}(t)}{\bar{F}(x)} dt, x < x_0.$$

Examples: normal law, exponential law, chi-square law, gamma law, lognormal law, Weibull distribution and laws with heavy queues decreasing to 0 faster than the exponential.

iii) The *Fréchet attraction domain* contains the distribution functions F so that $\bar{F}(x) = x^{-\alpha}L(x)$ with L slowly varying function:

$$\exists k > 0 : \lim_{x \to \infty} \frac{1 - F(x)}{1 - F(ux)} = u^k \text{, for all } u > 0. \quad [1.87]$$

Examples: Student's law, Cauchy's law, Pareto's law, laws with heavy queues decreasing to 0 slower than the exponential.

iv) The *Weibull attraction domain* contains the distribution functions F so that $\bar{F}(x_0 - x^{-1}) = x^{\alpha}L(x)$ with L being a slowly varying function.

Examples: distributions with bounded support (uniform distribution, Beta distribution, etc.).

1.3.12.3. *Exact values of the norming constants*

It is possible to obtain the exact values of the parameters $\Im_1$ (see EMB 08]). We give here some examples.

1) *Fréchet case*

$$c_n = \inf\left\{x \in \mathbb{R} : F(x) \geq 1 - \frac{1}{n}\right\}$$
$$(= F^{-1}(1 - \frac{1}{n}) \text{ for F strictly increasing}), \quad [1.88]$$

$$d_n = 0 \quad [1.89]$$

Examples:

Pareto-like distribution, stable with index $\alpha < 2$, Burr distribution:

$$F(x) \sim 1 - Kx^{-\alpha}, K, \alpha > 0 \Rightarrow \ d_n = 0, c_n = (Kn)^{\frac{1}{\alpha}}, \quad [1.90]$$

2) *Weibull case* (for $F : \exists x_0 > 0 : F(x_0) = 1$ et $F(x) < 1 \forall x < x_0$)

$$c_n = x_0 - \inf\left\{x \in \mathbb{R} : F(x) \geq 1 - \frac{1}{n}\right\}$$

$$(= F^{-1}(1 - \frac{1}{n}) \text{ for F strictly increasing}), \qquad [1.91]$$

$$\mathrm{d}_n = x_0.$$

Examples:

$$\text{Uniform distribution: } c_n = \frac{1}{n}, d_n = 1, \qquad [1.92]$$

3) *Gumbel case*

$$c_n = a(d_n)$$

$$d_n = \inf\left\{x \in \mathbb{R} : F(x) \geq 1 - \frac{1}{n}\right\}$$

$$(= F^{-1}(1 - \frac{1}{n}) \text{ for F strictly increasing}).$$

Examples:

$$\text{Exponential distribution of parameter } \lambda\text{: } d_n = \frac{\ln n}{\lambda}, c_n = \frac{1}{\lambda},$$

$$\text{reduced normal case: } c_n = \frac{1}{\sqrt{2\ln n}}, d_n = \sqrt{2\ln n} - \frac{\ln(4\pi) + \ln\ln n}{2\sqrt{2\ln n}} \qquad [1.93]$$

$$\text{lognormal } (\mu,\sigma)\text{: } c_n = \frac{\sigma}{\sqrt{2\ln n}} d_n, d_n = \mathrm{e}^{(\mu + \sigma(\sqrt{2\ln n} - \frac{\ln(4\pi)+\ln\ln n}{2\sqrt{2\ln n}}))}.$$

1.3.12.4. *Parameters estimation [ESC 97]*

Let us start with a sample of M risk observations restructured into N subgroups each with n observations where, of course, $M = n \times N$:

$$x_1, \ldots, x_n, x_{n+1}, \ldots, x_{2n}, \ldots, x_{(N-1)n}, \ldots, x_{Nn}.$$

For each of the N subgroups, we compute the maximum value:

$$y_{1,n} = \max\{x_1,...,x_n\},$$
$$y_{2,n} = \max\{x_{n+1},...,x_{2n}\}$$
$$\vdots$$
$$y_{N,n} = \max\{x_{(N-1)n+1},...,x_{Nn}\}.$$

This new sample is ranked in an increasing way:

$$y_1' \leq ... \leq y_N'.$$

Then it is possible to show that the parameters are a solution of a nonlinear regression problem:

$$-\ln\left[-\ln(\frac{i}{N+1})\right] = \frac{1}{\tau}(\ln\alpha_n - \ln\left[\alpha_n - \tau(y_i' - \beta_n)\right] + \varepsilon_i,$$
$$i = 1,..., N,$$
$$\varepsilon_1,...,\varepsilon_N \text{ independent and normal } N(0,\sigma^2).$$

For $\tau = 0$, we obtain a linear regression:

$$-\ln\left[-\ln(\frac{i}{N+1})\right] = \frac{y_i' - \beta_n}{\alpha_n}\frac{1}{\tau} + \varepsilon_i,$$
$$i = 1,..., N,$$
$$\varepsilon_1,...,\varepsilon_N \text{ independent et normal } N(0,\sigma^2).$$

REMARK 1.6.– The presented method is not frequently used as it has two main disadvantages:

i) the parameters are obtained by linear regression and so this estimation takes time;

ii) the number of observations needed for the linear regression must be large. For example, for $n = N = 50$, where N is the number of n realizations done, we need 2,500 observations; from this fact, it is not very easy to use daily data as we will need a historical record of 10 years, but of course with high-frequency trading (HFT) this critique is no longer true.

1.3.12.5. *Characteristic of extreme value distribution*

1) *Gumbel case*

From relations [1.43], we know that this distribution function is given by:

$$F_G = e^{-e^{-y}}. \qquad [1.94]$$

Let us also mention the following results:

$$\begin{aligned} &\frac{dF}{dx}(x) = f(x) = e^{(-x-e^{-x})}, \\ &E(X) = 0{,}57722..., \\ &\text{var}(X) = \frac{\pi^2}{6}, \\ &\gamma_1 = 1{,}29857, \gamma_2 = 5{,}4. \end{aligned} \qquad [1.95]$$

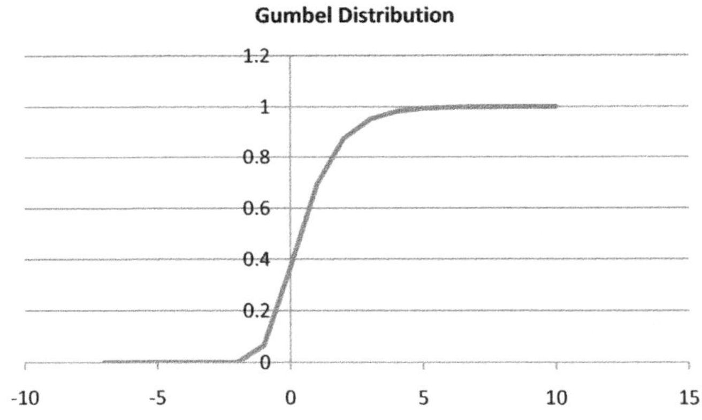

Figure 1.3. *Graph of the Gumbel distribution*

This is the normalized form, and the general form is given by:

$$\begin{aligned} &F_Y(x) = e^{-e^{-\frac{x-\mu}{\beta}}}, x \in \mathbb{R} \\ &(\mu \in \mathbb{R}, \beta > 0). \end{aligned} \qquad [1.96]$$

The values of the main parameters are given by:

$$E(Y) = \mu + \beta\gamma,$$
$$\sigma^2(Y) = \frac{\pi^2}{6}\beta^2 \qquad [1.97]$$

where γ is the Euler's constant defined by:

$$\gamma = \lim_{n \to \theta}((\sum_{k=1}^{n} \frac{1}{n}) - \ln n))$$

whose approximate value accurate to 10 decimal is γ = 0.5772156649.

2) *Fréchet case*

From relations [1.43], we know that this distribution function is given by:

$$F_\beta(x) = \begin{cases} 0, x \leq 0, \\ e^{-x^{-\beta}}, x > 0, \end{cases}, \beta > 0 \qquad [1.98]$$

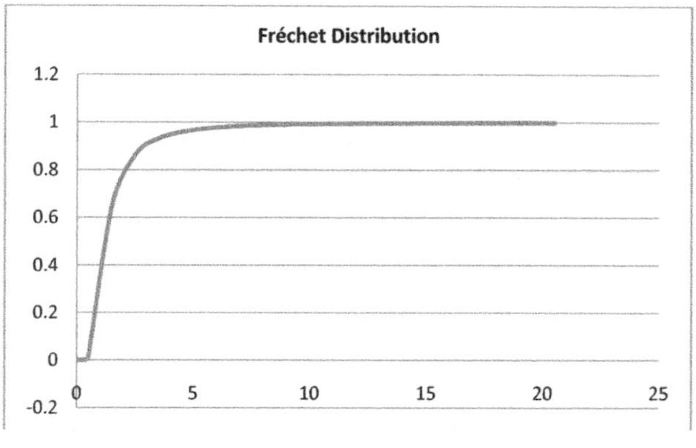

Figure 1.4. *Graph of the Fréchet distribution (β=2,1)*

and:

$$E(F) = \Gamma(1 - \frac{1}{\beta}), \beta > 1,$$
$$\sigma^2(F) = \Gamma(1 - \frac{2}{\beta}) - \Gamma^2(1 - \frac{1}{\beta}), \beta > 2. \qquad [1.99]$$

3) *Weibull case*

Here, we have:

$$F_{W,\beta}(x) = \begin{cases} e^{-(-x)^{\beta}}, & x < 0, \\ 1, & x \geq 0, \end{cases} \quad \beta > 0. \qquad [1.100]$$

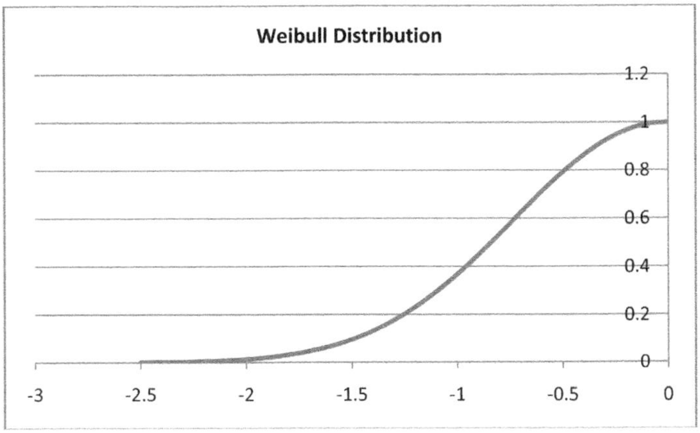

Figure 1.5. *Graph of the Weibull distribution (β=2,1)*

This is a distribution of a negative variable X and so $-X$ is a positive random variable having a standard Weibull distribution given by:

$$F_{-X}(x) = \begin{cases} e^{-(-x)^{\beta}}, & x \geq 0, \\ 0, & x < 0, \end{cases} \quad \beta > 0. \qquad [1.101]$$

With mean and variance given by:

$$m_{-X} = \Gamma(1+\frac{1}{\beta}), \sigma_{-X}^2 = \Gamma(1+\frac{2}{\beta}) - \Gamma^2(1+\frac{1}{\beta}). \qquad [1.102]$$

1.4. The normal power (NP) approximation

Let X be an insured risk of mean μ, variance σ^2 with $\gamma = \gamma_1$ different from zero.

The normal distribution even truncated is rarely adapted for measuring risks in non-life insurance essentially due to dissymmetry appearing on their distributions.

Nevertheless, in some cases of non-catastrophic risks, it is possible to take into account the dissymmetry to extend the normal distribution using the so-called NP approximation; this means that we can approximate well the distribution of X with the following relation:

$$P\left(\frac{X-\mu}{\sigma}\leq s+\frac{\gamma}{6}(s^2-1)\right)\approx\Phi(s), s\geq 1, \qquad [1.103]$$

or:

$$P\left(\frac{X-\mu}{\sigma}\leq x\right)\approx\Phi\left(\sqrt{\frac{9}{\gamma^2}+\frac{6x}{\gamma}+1}-\frac{3}{\gamma}\right), x\geq 1. \qquad [1.104]$$

EXAMPLE 1.1.–

To find the solvency capital requirement (SCR) related to the risk X that is the quartile $z_{0.995}$ such that:

$$P(X\leq z_{0,995})=0,995, \qquad [1.105]$$

or

$$P(\frac{X-\mu}{\sigma}\leq\frac{z_{0.995}-\mu}{\sigma})=0.995, \qquad [1.106]$$

we have from relation [1.103], with $s=\lambda_{0.995}$, the quartile at 99.5% of the standard normal distribution that:

$$P\left(\frac{X-\mu}{\sigma}\leq s+\frac{\gamma}{6}(s^2-1)\right)\approx 0.995, \qquad [1.107]$$

and so:

$$P\left(\frac{X-\mu}{\sigma}\leq s+\frac{\gamma}{6}(s^2-1)\right)\approx 0.995,$$

and consequently [1.108]

$$\frac{z_{0.995}-\mu}{\sigma}=\lambda_{0.995}+\frac{\gamma}{6}(\lambda_{0.995}^2-1).$$

The value of $z_{0,99}$ is thus given by:

$$z_{0,99} = \mu + \sigma[\lambda_{0,99} + \frac{\gamma}{6}(\lambda_{0,99}^2 - 1)]. \qquad [1.109]$$

By extension for the quartile z_α such that:

$$P(X \le z_\alpha) = \alpha, \qquad [1.110]$$

we obtain:

$$z_\alpha = \mu + \sigma[\lambda_\alpha + \frac{\gamma}{6}(\lambda_\alpha^2 - 1)]. \qquad [1.111]$$

Table 1.1 gives some examples of the impact of dissymmetry measured by γ on the values of $F(x,y) = P(X \le x, Y \le y)$. and λ_α by the value of the coefficient of σ in relation [1.111]:

Alpha	Lambda alpha	New quantile
0.990	2.33	2.33 + 0.74 * gamma
0.995	2.58	2.58 + 0.94 * gamma
0.999	3.09	3.09 + 1.43 * gamma
0.9995	3.29	3.29 + 1.64 * gamma
0.9999	3.72	3.72 + 2.14 * gamma

Table 1.1. *Impact of dissymmetry between z_α and λ_α*

Sometimes, it may be necessary to take into account the possibility of having large claims with probabilities larger than that in the normal case. This means that the considered risks may have heavy tails measured by the asymmetry coefficient or the kurtosis introduced in section 3.1, where γ_2 is complementary to the coefficient $\gamma = \gamma_1$. In this case, we use the *Cornish–Fisher* approximation, extending the result [1.111]:

$$z_\alpha = \mu + \sigma[\lambda_\alpha + \frac{\gamma}{6}(\lambda_\alpha^2 - 1) + \frac{1}{24}(\lambda_\alpha^3 - 3\lambda_\alpha)(\gamma_2 - 3) - \frac{1}{36}(2\lambda_\alpha^3 - 5\lambda_\alpha)\gamma^2]. \qquad [1.112]$$

1.5. Conditioning

The concept of conditioning is one of the most important factors for stochastic modeling as indeed it considers the impact of *supplementary information* on the considered probability distribution.

Let us begin to recall the well-known definition of the conditional probability of the event A given the event B:

$$P_B(A) = \frac{P(A \cap B)}{P(B)}, \forall A \in \Im . \qquad [1.113]$$

P_B is, in fact, a probability measure called the *conditional probability measure given B.*

The integral with respect to this measure is called the conditional expectation E_B relative to P_B.

From relation [1.12] and since $P_B(B) = 1$, we thus obtain for any integrable r.v. Y:

$$E_B(Y) = \int_\Omega Y(\omega) dP_B = \frac{1}{P(B)} \int_B Y(\omega) dP . \qquad [1.114]$$

We can now extend this definition to *arbitrary* sub-σ-algebras instead of the simple case of $\{\varnothing, B, B^c, \Omega\}$ using an extension of property [1.113] as a definition with the help of the Radon–Nikodym theorem (see [HAL 74]).

DEFINITION 1.8.– *If* $\Im_1$ *is a sub-σ-algebra of* $\Im$, *the conditional expectation of the integrable r.v. Y given* $\Im_1$, *denoted by* $E_{\Im_1}(Y)$ *or* $E(Y|\Im_1)$, *is any one r.v. of the equivalence class such that:*

i) $E_{\Im_1}(Y)$ is $\Im_1$-measurable;

ii) $\int_B E_{\Im_1}(Y)(\omega) dP = \int_B Y(\omega) dP, B \in \Im_1$. [1.115]

In fact, the class of equivalence contains all the random variables a.s. equally satisfying relation [1.114].

REMARK 1.7.– Taking $B=\Omega$ in relation [1.114], we obtain:

$$E(E_{\Im_1}Y)) = E(Y). \qquad [1.116]$$

Particular cases

i) $\Im_1$ *is* generated *by one r.v. X.*

This case means that $\Im_1$ is the sub-σ-algebra of $\Im$ generated by all the inverse images of X and we will use as notation:

$$E_{\Im_1}(Y) = E(Y|X), \qquad [1.117]$$

and this conditional expectation is called the *conditional expectation of Y given X.*

ii) $\mathbf{\Sigma} = \begin{bmatrix} \sigma_1^2 & \sigma_{12} \\ \sigma_{21} & \sigma_2^2 \end{bmatrix}$ is generated by n r.v. $X_1,...,X_n$.

This case means that $\Im_1$ is the sub-σ-algebra of $\Im$ generated by all the inverse images of $X_1,...,X_n$ and we will use as notation:

$$E_{\Im_1}(Y) = E(Y|X_1,...,X_n), \qquad [1.118]$$

and this conditional expectation is called the conditional expectation of Y given $X_1,...,X_n$.

In the latter case, it can be shown [LOÈ 77] that there exists a version $\varphi(X_1,...,X_n)$ of the conditional expectation so that φ is a Borel function from $\mathbb{R}^n$ to $\mathbb{R}$ and as such it follows that $E(Y|X_1,...,X_n)$ is constant on each set belonging to $\Im_1$ for which $X_1(\omega)=x_1,...,X_n(\omega)=x_n$, for instance.

This justifies the following abuse of notation:

$$E\left(Y|X_1(\omega)=x_1,...,X_n(\omega)=x_n\right) = \varphi(x_1,...,x_n) \qquad [1.119]$$

representing the value of this conditional expectation on all the ω's belonging to the set $\{\omega : X_1(\omega)=x_1,...,X_n(\omega)=x_n\}$.

Taking $B = \Omega$ in relation [1.115], we obtain:

$$E(Y) = \int_{R_n} E\left(Y \middle| X_1(\omega) = x_1, ..., X_n(\omega) = x_n\right) dP(X_1(\omega) \leq x_1, ..., X_n(\omega) \leq x_n) \quad [1.120]$$

a result often used in the sequel to evaluate the mean of a random variable using its conditional expectation with respect to some given event.

iii) If $\Im_1 = \{\varnothing, \Omega\}$, we get $E(Y|\Im_1) = E(Y)$, and if $\Im_1 = \{\varnothing, B, B^c, \Omega\}$, then $E(Y|\Im_1) = E(Y|B)$ on B and $E(Y|\Im_1) = E(Y|B^c)$ on B^c.

iv) Taking as r.v. Y the indicator of the event A, that is to say:

$$1_{A(\omega)} = \begin{cases} 1, \omega \in A, \\ 0, \omega \notin A, \end{cases} \quad [1.121]$$

the conditional expectation becomes the *conditional probability of A given* $\Im_1$ and is denoted as follows:

$$P(A|\Im_1) = E(1_A(\omega)|\Im_1) \quad [1.122]$$

and then relation [1.114] becomes:

$$\int_B P\left(A|\Im_1(\omega)\right) dP = P(A \cap B), B \in \Im_1. \quad [1.123]$$

Letting $B = \Omega$ in this final relation, we obtain:

$$E\left(P\left(A|\Im_1\right)\right) = P(A), \quad [1.124]$$

a property extending the theorem of total probability.

If, moreover, A is independent of $\Im_1$, that is to say, if for all B belonging to $\Im_1$:

$$P(A \cap B) = P(A)P(B), \quad [1.125]$$

then we see from [1.124] that:

$$P\left(A|\Im_1\right)(\omega) = P(A), \omega \in \Omega. \quad [1.126]$$

Similarly, if the r.v. Y is independent of $\Im_1$, that is to say if for each event B belonging to $\Im_1$ and each set A belonging to the σ-algebra generated by the inverse images of Y, denoted by σ (Y), we have:

$$E\left(Y\middle|\Im_1\right)=E(Y). \tag{1.127}$$

Indeed, from relation [1.114], we can write that:

$$\begin{aligned}\int_B E_{\Im_1}(Y)(\omega)dP &= \int_B Y(\omega)dP, B\in\Im_1,\\ &= E\left(Y1_B\right),\\ &= E(Y)P(B),\\ &= \int_B E(Y)dP,\end{aligned} \tag{1.128}$$

and so, relation [1.127] is proved.

In particular, if $\Im_1$ is generated by the r.v. $X_1,\dots,X_n$, then the independence between Y and $\Im_1$ implies that:

$$E\left(Y\middle|X_1,\dots,X_n\right)=E(Y). \tag{1.129}$$

Relations [1.126] and [1.128] allow us to have a better understanding of the *intuitive meaning of conditioning* and its importance in stochastic modeling.

Under independence assumptions, conditioning has absolutely no impact, for example, on the expectation or the probability; on the contrary, dependence implies that the results with or without conditioning will be different; this fact means that we can interpret conditioning as given *additional information* useful to get more precise results in the case of dependence, of course of an asset.

The properties of expectation, quoted in section 1.2, are also properties of conditional expectation, true now a.s., but there are supplementary properties which are very important in stochastic modeling. They are given in the following proposition:

PROPOSITION 1.1 (SUPPLEMENTARY PROPERTIES OF CONDITIONAL EXPECTATION).– On the probability space $(\Omega,\Im,P)$, we have the following properties:

i) If the r.v. X is $\Im_1$-measurable, then:

$$E(X|\Im_1)=X, a.s. \tag{1.130}$$

ii) Let X be a r.v. and Y be $\Im_1$ -measurable, then:

$$E(XY|\Im_1) = YE(X|\Im_1), a.s. \tag{1.131}$$

To obtain the last but most important property, we have from relation [1.115] that $E_\Im(Y) = Y$, and so taking $Y = E_{\Im_1}(Y)$, we see that:

$$E_\Im(E_{\Im_1}(Y)) = E_{\Im_1}(Y) \tag{1.132}$$

and of course since:

$$E_{\Im_1}(E_\Im(Y)) = E_{\Im_1}(Y). \tag{1.133}$$

Putting these last two relations together, we obtain:

$$E_\Im(E_{\Im_1}(Y)) = E_{\Im_1}E_\Im((Y)) = E_{\Im_1}(Y). \tag{1.134}$$

This last result may be generalized now taking a sub-σ-algebra $\Im_2$ instead of $\Im$ and such that $\Im_1 \subset \Im_2$ so giving the important following result.

PROPOSITION 1.2 (SMOOTHING PROPERTY OF CONDITIONAL EXPECTATION).–

Let $\Im_1$, $\Im_2$ be two sub-σ-algebras of $\Im$ such that $\Im_1 \subset \Im_2$; then it is true that:

$$E_{\Im_2}(E_{\Im_1}(Y)) = E_{\Im_1}(E_{\Im_2}(Y)) = E_{\Im_1}(Y), \tag{1.135}$$

a property called the smoothing property in [LOÈ 77].

A particular case of [1.135] is, for example, that:

$$E\left(E\left(Y|X_1,...,X_n\right)|X_1\right) = E\left(E\left(Y|X_1\right)|X_1,...,X_n\right) = E\left(Y|X_1\right). \tag{1.136}$$

Here is an example illustrating this interest for sums of a random number of random variables with the so-called *Wald identities*.

EXAMPLE 1.2 (WALD'S IDENTITIES).– Let $(X_n, n \geq 1)$ be a sequence of i.i.d. real random variables and N be a non-negative r.v. with integer values independent of the given sequence. The random variable defined by:

$$S_N = \sum_{n=1}^{N} X_n \quad [1.137]$$

is called a *sum of a random number of random variables* and the problem to be solved is the computation of the mean and the variance of this sum assuming that the r.v. X_n have a variance.

The two Wald's identities (see [JAN 09a] for a proof) are successively given by:

$$E(S_N) = E(N)E(X), \quad [1.138]$$

$$\operatorname{var}(S_N) = E(N)\operatorname{var}(X) + \operatorname{var}(N)(E(X))^2 \quad [1.139]$$

REMARK 1.8.– In the particular case of an n-dimensional real r.v. $X = (X_1,...,X_n)$, we can now introduce the very useful definition of the *conditional distribution function of X given* $\Im_1$ defined as follows:

$$\begin{aligned} F_{\Im_1}(x_1,...,x_n,\omega) &= P\left(X_1 \leq x_1,...,X_n \leq x_n \,|\, \Im_1\right) \\ &= P\left(\{\omega' : X_1(\omega') \leq x_1,...,X_n(\omega') \leq x_n\} \,|\, \Im_1)(\omega)\right). \end{aligned} \quad [1.140]$$

Another useful definition concerns an extension of the concept of the independence of random variables to the definition of *conditional independence of the n variables* $X_1,\ldots,X_n$. These *n r.v.* are *conditional independent* if for all $(x_1,\ldots,x_n)$ belonging to $\mathbb{R}^n$ we have the following identity:

$$F_{\Im_1}(x_1,...,x_n,\omega) = \prod_{k=1}^{n} F_{\Im}(x_k,\omega), \quad [1.141]$$

where, of course, we have:

$$F_{\Im}(x_k,\omega) = P\left(X_k \leq x_k \,|\, \Im_1\right). \quad [1.142]$$

EXAMPLE 1.3.– Let (X, Y) be a two-dimensional real r.v. whose d.f. is given by:

$$F(x,y) = P(X \leq x, Y \leq y). \quad [1.143]$$

As $\mathbb{R}^2$ is a complete separable metric space, it is previously mentioned that regular conditional probabilities exist given the sub-σ-algebras $\sigma(X)$ or $\sigma(Y)$, and so the related conditional d.f. denoted by:

$$F_{X|Y}\left(x\middle|\{\omega : Y = y\}\right),\ F_{Y|X}\left(y\middle|\{\omega : X = x\}\right) \qquad [1.144]$$

also exists.

If, moreover, the d.f. F has a density f, we can also introduce the concept of *conditional density* for both functions $F_{X|Y}$, $F_{Y|X}$, F_X, giving at the same time an intuitive interpretation of conditioning in this special case.

We know that for every fixed (x,y):

$$f(x,y)\Delta x\Delta y + o(x,y,\Delta x,\Delta y) = P(x < X \leq x + \Delta x, y < Y \leq y + \Delta y), \qquad [1.145]$$

where $o(x,y,\Delta x,\Delta y) \to 0$ for $(\Delta x,\Delta y) \to (0,0)$, and similarly for the marginal density function of X:

$$f_X(x)\Delta x + \bar{o}(x,\Delta x) = P(x < X \leq x + \Delta x), \qquad [1.146]$$

where $\bar{o}(x,\Delta x) \to 0$ for $\Delta x \to 0$ with of course:

$$f_X(x) = \int_R f(x,y)dy. \qquad [1.147]$$

We thus obtain:

$$P\left(y < Y \leq y + \Delta y \middle| x < X \leq x + \Delta x\right) = \frac{f(x,y)\Delta x\Delta y + o(x,y,\Delta x,\Delta y)}{f_X(x)\Delta x + \bar{o}(x,\Delta x)}. \qquad [1.148]$$

Letting Δx tend to 0, we obtain:

$$\lim_{\Delta x \to 0} P\left(y < Y \leq y + \Delta y \middle| x < X \leq x + \Delta x\right) = \frac{f(x,y)}{f_X(x)}\Delta y. \qquad [1.149]$$

This relation shows that the function $f_{Y|X}$ is defined by:

$$f_{Y|X}\left(y\middle|x\right) = \frac{f(x,y)}{f_X(x)} \qquad [1.150]$$

which is, by defintion, the conditional density function of Y, given X. Similarly, the conditional density of X, given Y, is given by:

$$f_{X|Y}(x|y) = \frac{f(x,y)}{f_Y(y)}. \qquad [1.151]$$

Consequently, for any Borel subsets A and B of $\mathbb{R}$, we have:

$$\begin{aligned} P(X \in A|Y(\omega) = y) &= \int_A f_{X|Y}(x|y)dx = \frac{1}{f_Y(y)}\int_A f(x,y)dx, \\ P((X,Y) \in A \cap B) &= \int_{A \cap B} f(x,y)dxdy = \int_B\left(\int_A f_{X|Y}(x|y)dx\right)f_Y(y)dy. \end{aligned} \qquad [1.152]$$

The last equalities show that the density of (X,Y) can also be characterized by one marginal d.f. and the associated conditional density, as from relations [1.103] and [1.10]:

$$f = f_X \times f_{Y|X} = f_Y \times f_{X|Y}. \qquad [1.153]$$

It is possible that *conditional means* exist; if so, they are given by the following relations:

$$E(X|Y=y) = \int_{\mathbb{R}} f(x|y)dx, \; E(Y|X=x) = \int_{\mathbb{R}} f(y|x)dy. \qquad [1.154]$$

The conditional mean of X (respectively Y) given $Y = y$ (respectively $X = x$) can be seen as a function of the real variable y (respectively x), called *the regression curve of* X (respectively Y) given Y (respectively X).

The two regression curves will generally not coincide and not be straight lines except if the two r.v. X and Y are independent because, in this case, we have from relations [1.150] and [1.27) that:

$$f_{X|Y} = f_X, f_{Y|X} = f_Y \qquad [1.155]$$

and so:

$$E(X|Y) = E(X), \; E(Y|X) = E(Y), \qquad [1.156]$$

proving thus that the two regression curves are straight lines parallel to the axes passing through the "center of gravity" ($E(X)$, $E(Y)$) of the probability mass in $\mathbb{R}^2$.

In the special case of a non-degenerated normal distribution for (X,Y) with vector mean (m_1,m_2) and variance–covariance matrix:

$$\mathbf{\Sigma}=\begin{bmatrix}\sigma_1^2 & \sigma_{12}\\ \sigma_{21} & \sigma_2^2\end{bmatrix}, \qquad [1.157]$$

it can be shown that the two conditional distributions are also normal with parameters:

$$\begin{aligned} Y\left|X\right. &\prec N_2\left(\mu_2+\rho\frac{\sigma_2}{\sigma_1}(x-\mu_1),\sigma_2^2(1-\rho^2)\right),\\ X\left|Y\right. &\prec N_2\left(\mu_1+\frac{1}{\rho}\frac{\sigma_1}{\sigma_2}(y-\mu_2),\sigma_2^2(1-\rho^2)\right). \end{aligned} \qquad [1.158]$$

1.6. Stochastic processes

In this section, we will always consider a *complete* probability space $(\Omega,\Im,\mathrm{P})$ with a *filtration F*.

Let us recall that a probability space $(\Omega,\Im,\mathrm{P})$ is *complete* if every subset of an event of probability 0 is measurable, i.e. in the σ-algebra $\Im$, and so also of probability 0.

DEFINITION 1.9.– *F is a filtration on the considered basic probability space if F is a family of* $(\Im_t, t\in T)$ *of sub-σ-algebras of* $\Im$*, the index set T being either the natural set* $\{0,1,\ldots,n,\ldots\}$ *or the positive half real line* $[0,\infty)$ *such that:*

$$\begin{aligned} &\text{i) } s<t\Rightarrow \Im_s\subset\Im_t,\\ &\text{ii) } \Im_t=\bigcap_{u>t}\Im_u, \end{aligned} \qquad [1.159]$$

iii) $\Im_0$ *contains all subsets with probability* 0.

Assumption (ii) is called the *right continuity property* of the filtration *F*.

Any filtration satisfying these three assumptions is called a filtration *satisfying the usual assumptions*.

The concept of filtration can be interpreted as a family of amounts of information so that $\Im_t$ gives all the observable events at time t.

DEFINITION 1.10.– *The quadruplet* $((\Omega, \Im, \mathrm{P}, (\Im_t, t \in T))$ *is called a filtered probability space.*

DEFINITION 1.11.– *A random variable* $\tau : \Omega \mapsto T$ *is a stopping time if:*

$$\forall t \in T : \ \{\omega : \tau(\omega) \leq t\} \in \Im_t. \qquad [1.160]$$

The interpretation is as follows: the available information at time t gives the possibility of observing the event given in [1.158] and to decide, for example, whether one stops the future observations after time t or not.

We have the following proposition:

PROPOSITION 1.3.– The random variable τ is a stopping time if and only if

$$\{\omega : \tau(\omega) < t\} \in \Im_t, \ \forall t \in T. \qquad [1.161]$$

DEFINITION 1.12.– *A stochastic process (or simply process) with values in the measurable space* $(E, \aleph)$ *is a family of random variables:*

$$\{X_t, t \in T\} \qquad [1.162]$$

where for all t:

$$X_t : \Omega \mapsto E, \ (\Im, \aleph) - measurable.$$

This means, in particular, that for every subset B of the σ-algebra $\aleph$, the set

$$X_t^{-1}(B) = \{\omega : X_t(\omega) \in B\} \qquad [1.163]$$

belongs to the σ-algebra $\Im$.

REMARK 1.9.– If $(E,\aleph)=(\mathbb{R},\beta)$, the process is called a *real stochastic process* with values in $\mathbb{R}$; if $(E,\aleph)=(\mathbb{R}^n,\beta^n)$, it is called a *real multi-dimensional process* with values in $\mathbb{R}^n$.

If T is the natural set $\{0,1,...,n,...\}$, the process X is called a *discrete-time stochastic process* or a *random sequence*; if T is the positive half real line $[0,\infty)$, the process X is called a *continuous-time stochastic process.*

DEFINITION 1.13.– *The stochastic process X is adapted to the given filtration if, for all t, the r.v.* X_t *is* $\Im_t$ *-measurable. This means that for all* $t\in T$ *:*

$$X_t^{-1}(B)=\{\omega: X_t(\omega)\in B\}\in\Im_t,\forall B\in\aleph. \quad [1.164]$$

DEFINITION 1.14.– *Two processes x and y are indistinguishable if a.s. for all* $t\in T$ *:*

$$X_t=Y_t. \quad [1.165]$$

This means that

$$\forall t: \mathrm{P}(X_t=Y_t)=1.$$

DEFINITION 1.15.– *The process X (or Y) is a modification of the process Y (or X) if a.s. for all* $t\in T$ *:*

$$X_t=Y_t.\ a.s. \quad [1.166]$$

This means that:

$$\mathrm{P}(X_t=Y_t.,\forall t\in T)=1. \quad [1.167]$$

for all $t\in T$.

DEFINITION 1.16.– *For every stochastic process x, the function from t to e*

$$t\mapsto X_t(\omega) \quad [1.168]$$

defined for each $\omega\in\Omega$, is called a *trajectory* or *sample path* of the process.

These sample paths represent simulation with a model to construct what is now called a *scenario generator.*

It must be clearly understood that the so-called "modern" study of stochastic processes is concerned with the study of the properties of these trajectories.

For example, we can affirm that if two processes X and Y are indistinguishable, then there exists a set N belonging to $\Im$ of probability 0 such that:

$$\forall \omega \notin N : X_t(\omega) = Y_t(\omega), \forall t \in T. \quad [1.169]$$

In other words, for each ω element of the set $\Omega - N$, the two functions $t \mapsto X_t(\omega)$ and $t \mapsto Y_t(\omega)$ are equal.

As the basic probability space is complete, the neglected set N belongs to $\Im_t$, for all $t \in T$.

DEFINITION 1.17.– *A real stochastic process X is càdlàg if a.s. the trajectories of X are right continuous and have left limits t for all* $t \in [0,T]$.

DEFINITION 1.18.– *If X is a real stochastic process and a set* $\Lambda \in \beta$*, then the r.v. defined by:*

$$T(\omega) = \inf\{t > 0 : X_t(\omega) \in \Lambda\} \quad [1.170]$$

is called the *hitting time* of Λ by the process X.

It is easily shown that the *properties* of stopping and hitting times are (see [PRO 90]):

i) If X is càdlàg, adapted and $\Lambda \in \beta$, then the hitting time related to Λ is a stopping time.

ii) Let S and T be two stopping times, then the following r.v.:

$$S \wedge T\left(= \min\{S,T\}\right),\ S \vee T\left(= \max\{S,T\}\right),\ S + T,\ \alpha S(\alpha > 1) \quad [1.171]$$

are also stopping times.

DEFINITION 1.19.– *If T is a stopping time, the* σ *-algebra* $\Im_T$ *defined by:*

$$\Im_T = \left\{\Lambda \in \Im : \Lambda \cap \{\omega : T(\omega) \le t\} \in \Im_t, \forall t \ge 0\right\} \quad [1.172]$$

is called the stopping time σ-algebra.

In fact, the σ-algebra $\Im_T$ represents the information of all observable sets up to the stopping time T. We can also say that $\Im_T$ is the smallest stopping time containing all the events related to the r.v. $X_{T(\omega)}(\omega)$ for all the adapted càdlàg processes X or generated by these r.v.

We also have for two stopping times S and T:

i) $S \leq T$ a.s. $\Rightarrow \Im_S \subset \Im_T$, [1.173]

ii) $\Im_S \cap \Im_T, = \Im_{S \wedge T}$. [1.174]

1.7. Martingales

In this section, we will briefly present some topics related to the most well-known category of stochastic processes, called *martingales*.

Let X be a real stochastic process defined on the filtered complete probability space $(\Omega, \Im, P, (\Im_t, t \in T))$.

DEFINITION 1.20.– *The process X is called a* $(\Im_t)$ *-martingale if:*

i) $\forall t \geq 0, \exists E(X_t)$, [1.175]

ii) $s < t \Rightarrow E(X_t \mid \Im_s) = X_s$, a.s. [1.176]

The latter equality is called the *martingale property* or the *martingale equality*.

DEFINITION 1.21.– *The process X is called a supermartingale (respectively submartingale) if:*

$$\forall t \geq 0, \exists E(X_t),$$

and

$$s < t \Rightarrow E(X_t \mid \Im_s) \leq (\geq) X_s, \text{a.s.}$$

The martingale concept is interesting; indeed, as the best estimator at time s (s>t) for the value of X_t, as given by the conditional expectation appearing in relation [1.176], the martingale equality means that *the best predicted value* is simply the observed value of the process at the time of predicting s.

The use of martingale in finance is frequently used to model the concept of an *efficient financial market.*

DEFINITION 1.22.– *The martingale X is closed if:*

$\exists Y$:
i) $\mathrm{E}(|Y|) < \infty$,
ii) $\forall t \in [0,\infty): \mathrm{E}(Y|\Im_t) = X_t$, a.s..

It is possible to prove the following result (e.g. see [PRO 90]):

PROPOSITION 1.4.–

i) If X is a supermartingale, then the function $t \mapsto E(X_t)$ is right continuous iff a unique modification Y of X exists such that Y is càdlàg.

ii) If X is a martingale, then, up to a modification, the function $t \mapsto E(X_t)$ is right continuous.

It follows that every martingale such that the function $t \mapsto E(X_t)$ is right continuous is càdlàg.

The two most important results concerning martingales are the martingale convergence theorem and the optional sampling (or Doob's) theorem.

Before giving these results, we still need a final technical definition.

DEFINITION 1.23.– [MEY 66]

A family $(\xi_u, u \in A)$, where A is an infinite index set, is *uniformly integrable* if:

$$\lim_{n\to\infty} \sup_{\alpha} \int_{\{\omega | |\xi_\alpha(\omega)| \geq n\}} |\xi_\alpha(\omega)| \, d\mathrm{P}(\omega) = 0 . \qquad [1.177]$$

PROPOSITION 1.5.– Let x be a supermartingale in such a way that the function $t \mapsto E(X_t)$ is right continuous such that:

$$\sup_{t\in[0,\infty)} E(|X_t|) < \infty ; \qquad [1.178]$$

then there exists a r.v. Y such that:

$$\begin{aligned}&\text{i) } E(|Y|),\\&\text{ii) } Y=\lim_{t\to\infty}X_t,\quad \text{a.s.}\end{aligned}\qquad [1.179]$$

Moreover, if X is a martingale closed by the r.v. Z, then the r.v. Y also closes X and:

$$Y=E\left(Z\middle|\mathfrak{I}_\infty\right),\qquad [1.180]$$

where

$$\mathfrak{I}_\infty=\sigma\left(\bigcup_{0\le t<\infty}\mathfrak{I}_t\right).\qquad [1.181]$$

With the help of the concept of uniform integrability, we can obtain the following corollary:

COROLLARY 1.1.–

i) Let X be a right continuous martingale and uniformly integrable; then there exists a.s. the following limit:

$$Y=\lim_{t\to\infty}X_t\qquad [1.182]$$

Moreover, $Y\in L^1$ and the r.v. Y closes the martingale X.

ii) Let X be a right continuous martingale; then $X=(X_t,t\ge 0)$ is uniformly integrable if and only if:

$$Y=\lim_{t\to\infty}X_t\qquad [1.183]$$

exists a.s., $Y\in L^1$, and $\left(X_t,t\in[0,\infty]\right)$ is a martingale with a.s.:

$$X_\infty=Y.\qquad [1.184]$$

Now, an interesting question is: what happens if we observe a martingale X at two stopping times S,T ($S<T$, a.s.)? The answer is given by the so-called *optional sampling theorem*, also called *Doob's theorem*.

PROPOSITION 1.6 (THE OPTIONAL SAMPLING THEOREM OR DOOB'S THEOREM).– Let X be a right continuous martingale closed by X_∞ and let S and T be two stopping times so that a.s. $S < T$; then, the r.v. $X_S, X_T \in L^1$ and:

$$X_S = E\left(X_T \mid \Im_S\right), \text{ a.s.} \quad [1.185]$$

This important theorem means that if we restrict the random observation time set to $\{S,T\}$, then the restriction of the martingale to this set is *still* a martingale provided that S and T are two stopping times with $S<T$, a.s.

This result is interesting for the concept of *stopped process.*

DEFINITION 1.24.– *Let X be a stochastic process and T be a stopping time. The stopped stochastic process X^T is defined by:*

$$X^T = \left(X_t^T, t \in [0,\infty]\right) \quad [1.186]$$

where

$$\begin{aligned} &X_t^T(\omega) = X_{t\wedge T}(\omega), \\ &\text{with } t \wedge T = \inf\{t,T\}. \end{aligned} \quad [1.187]$$

From this definition, it follows that if the process X is adapted and càdlàg, then so is the stopped process X^T. This is due to the fact that $t \wedge T$ is also a stopping time and moreover:

$$X_t^T = X_t 1_{\{t<T\}} + X_T 1_{\{t \geq T\}}. \quad [1.188]$$

This leads to the last result we want to mention.

PROPOSITION 1.7.– Let X be a right continuous uniformly integrable martingale; then the stopped process $X^T = \left(X_{t\wedge T}, t \in [0,\infty]\right)$ has the same properties with respect to the filtration $\left(\Im_t, t \in [0,\infty]\right)$.

2

Homogeneous and Non-homogeneous Renewal Models

In this chapter, we will present both homogeneous and non-homogeneous renewal processes, though the latter are not quite established. One of the tools necessary for studying these processes is the non-homogeneous time convolution.

Classical renewal theory has a great relevance in general in economics and in particular in actuarial science. However, in most actuarial problems, the age of the insured person has a large influence on obtaining more realistic models and therefore we need non-homogeneous renewal theory. We will also develop the numerical aspects of these processes and treat a real data application to an aspect of motorcar insurance.

2.1. Introduction

Renewal theory was introduced a long time ago in a paper by Lotka [LOT 40] concerning industrial replacement. The theory was fully studied by Feller [FEL 41], Doob [DOO 48] and Smith [SMI 54, SMI 58]. Nowadays, this theory is presented as a standard theory in most textbooks on stochastic processes, but essentially in the case of time homogeneity. In a homogeneous renewal process, the system is renewed when the studied phenomenon is verified, then it restarts with the same initial characteristics. In the non-homogeneous case, the times of renewal influence the properties of the system.

This is particularly useful for most actuarial models for which the time variable taken into account is age.

Let us nevertheless point out that if many researchers have worked with non-homogeneous Poisson's processes, it is no more the case with general non-homogeneous processes.

The non-homogeneous Poisson's process is a particular case of the non-homogeneous renewal process, but its application is limited as the waiting time distribution function (d.f.), between the two renewals, must be negative exponentials. This constraint, as will be seen from our application, is usually not satisfied in real-world problems.

To extend theoretical results from homogeneous to non-homogeneous renewal processes, it is necessary to introduce what we call non-homogeneous time convolutions (see [GIS 15]).

Regarding non-homogeneous convolutions, it should be noted that this topic seems to be new as well as their application of general non-homogeneous renewal processes in actuarial science.

If the cumulating d.f. of the renewal process is of a negative exponential type, then the renewal process becomes a Poisson's process (see [JAN 06] and [MIK 09]) and the related integral equation, both in homogeneous and non-homogeneous cases, can be solved analytically [ÇIN 75]. This is a very particular case. In a more general case, the renewal equation cannot usually be solved analytically and it is necessary to solve it numerically.

The chapter is organized as follows. In section 2.2, the continuous time non-homogeneous convolutions and their properties are presented for the first time. Section 2.3 describes both continuous time homogeneous and non-homogeneous renewal processes and their evolution equation. In section 2.4, the numerical solution of the evolution equation of the non-homogeneous continuous time renewal process is presented. Furthermore, we show how to obtain the discrete time evolution equation from the continuous time evolution equation and vice versa. In section 2.5, an application to motorcar insurance in both the homogeneous and non-homogeneous environment is reported. Section 2.6 presents some short conclusive remarks.

Let us also say that as transient time models are the one that is useful for non-life insurance, we will not fully develop asymptotical results for renewal theory. They can be found for example in [COX 65] and [JAN 09a].

2.2. Continuous time non-homogeneous convolutions

2.2.1. *Non-homogeneous convolution product*

Let us begin by saying that all considered functions are of course measurable functions of the *ad hoc* space in $\mathbb{R}$ or $\mathbb{R}^+$.

DEFINITION 2.1.– *A measurable function of* $\mathbb{R}^2 \mapsto \mathbb{R}^+$ $f(s,t), 0 \leq s \leq t$, *where s, t represent times, is time non-homogeneous if:*

$$\exists (s,t) \neq (s',t'), t-s = t'-s' : f(s,t) \neq f(s',t') \qquad [2.1]$$

DEFINITION 2.2.– *i) Given two time integrable non-negative non-homogeneous functions* $f(s,t), g(s,t)$ *their convolution is defined in the following way:*

$$(f * g)(s,t) = \int_s^t g(s,\tau) f(\tau,t) d\tau; \; s,t \in \mathbb{R}, s < t. \qquad [2.2]$$

ii) Let F and G be measurable or Borel functions of two variables such that for all s, F(s,.) and G(s,.) are distribution functions on $\inf\{X,Y\}, \sup\{X,Y\}, X+Y, X-Y, X \cdot Y, \frac{X}{Y}$ *and such that for all t, F(.,t) and G(.,t) are continuous, then their convolution product is defined as follows:*

$$(F \otimes G)(s,t) = \int_s^t F(s,\tau) dG(\tau,t), \; s,t \in \mathbb{R}. \qquad [2.3]$$

PROPOSITION 2.1.– The non-homogeneous convolution product is associative, distributive and bi-linear.

PROOF.– We only give the proof for definition [2.2], the proof being similar with definition [2.3].

i) *Associativity*: given the three functions of two variables *(s,t)* f, g, h, we have to prove that:

$$(f * (g * h))(s,t) = ((f * g) * h)(s,t)$$

Using Fubini's theorem, the following successive transformations lead to the result:

$$\begin{aligned}(f*(g*h))(s,t) &= f*\left(\int_s^t h(s,\tau_1)g(\tau_1,t)d\tau_1\right)\\ &= \int_s^t\left(\int_s^{\tau_2} h(s,\tau_1)g(\tau_1,\tau_2)d\tau_1\right)f(\tau_2,t)d\tau_2\\ &= \int_s^t\int_s^{\tau_2}(h(s,\tau_1)g(\tau_1,\tau_2))f(\tau_2,t)d\tau_1 d\tau_2\\ &= \int_s^t\int_{\tau_1}^t h(s,\tau_1)g(\tau_1,\tau_2)f(\tau_2,t)d\tau_2 d\tau_1\\ &= \int_s^t h(s,\tau_1)\int_{\tau_1}^t g(\tau_1,\tau_2)f(\tau_2,t)d\tau_2 d\tau_1\\ &= \int_s^t h(s,\tau_1)(f*g(\tau_1,t))d\tau_1 = ((f*g)*h)(s,t).\end{aligned}$$

Distributivity: given the three functions of two variables (s,t) f,g,h, we have to prove that:

$$(f*(g+h))(s,t) = (f*g)(s,t)+(f*h)(s,t)\text{. Left distributivity}$$

$$((f+g)*h)(s,t) = (f*h)(s,t)+(g*h)(s,t)\text{. Right distributivity}$$

PROOF.– Let us prove only the left distributivity. Indeed, we successively have:

$$\begin{aligned}&(f*(g+h))(s,t)\\ &= \int_s^t (g+h)(s,\tau)f(\tau,t)d\tau = \int_s^t (g(s,\tau)f(\tau,t)+h(s,\tau)f(\tau,t))d\tau\\ &= (f*g)(s,t)+(f*h)(s,t)\end{aligned}$$

Bi-linearity: the additivity is ensured by the distributivity. It remains to prove that $\forall a\in\mathbb{R}$, $\forall f,g$ it results:

$$(a(f*g))(s,t) = ((af)*g)(s,t) = (f*(ag))(s,t).$$

PROOF. –

$$(a(f*g)(s,t)) = a\int_s^t g(s,\tau)f(\tau,t)d\tau = \int_s^t g(s,\tau)af(\tau,t)d\tau = ((af)*g)(s,t)$$

$$a\int_s^t g(s,\tau)f(\tau,t)d\tau = \int_s^t ag(s,\tau)f(\tau,t)d\tau = (f*(a\,g))(s,t).$$ □

REMARK 2.1.– Though the convolution product in the homogeneous case is commutative for distribution and density functions, it is no longer the case for the non-homogeneous case i.e.:

$$\exists f, g \text{ functions of } (s,t): (f*g)(s,t) \neq (g*f)(s,t).$$

It can be verified that with the functions $f(s,t) = e^{3s+4t}, g(s,t) = e^{-4s+2t}$, their convolution is not commutative.

REMARK 2.2.– *Time homogeneity* is a particular case of non-homogeneity. Indeed, by definition here, a function of two real variables is *homogeneous* if:

$$f(s,t) = f(s+h, t+h);\ \forall 0 \leq s \leq t \text{ and } h: -s < h.$$

Given this hypothesis, it is possible to introduce a new function but now of one variable called $\hat{f}$ and is defined by:

$$\hat{f}(\tau) = f(s,t),\ \forall s,t \text{ and } \tau = t-s. \qquad [2.4]$$

s and *t* being values of time with $s < t$, representing the starting and arriving times of a given event such as a stay in a given state of a physical system, it results that $\hat{f}$ is a function of the time difference that can be interpreted as the time *t-s* already spent in this time and not of the starting and arriving times.

REMARK 2.3.– i) Let us consider the functions *f* and *g* of (*s,t*) homogeneous in time. Let:

$$\hat{f}(x) = f(t-s) \text{ and } \hat{g}(x) = g(t-s), x = t-s,$$

then from the homogeneity hypothesis and, it results:

$$(f*g)(s,t)=\int_s^t g(s,\tau)f(\tau,t)d\tau=\int_s^t \hat{g}(\tau-s)\hat{f}(t-\tau)d\tau$$

$$=\int_0^u \hat{g}(x)\hat{f}(u-x)dx=(\hat{f}*\hat{g})(u),\ \text{where } u=t-s$$

$$P\left(\{\omega:\lim X_n(\omega)=X(\omega)\}\right)=1$$

In fact, this latter formula gives the definition of the *classical convolution product* of two non-negative measurable functions $f,g:\mathbb{R}^+\mapsto\mathbb{R}^+$.

We have already said that if the two functions are densities, then their convolution product is commutative.

$$\int_0^t g(x)f(t-x)dx=(f*g)(t),t\in\mathbb{R}^+. \qquad [2.5]$$

If $f = g$, by the associativity property, we can define successive convolutions products of a non-negative functions f as follows:

$$f^{(2)}(t)=\int_0^t f(x)f(t-x)dx,$$

$$\vdots \qquad [2.6]$$

$$f^{(n)}(t)=\int_0^t f(x)f^{(n-1}(t-x)dx.$$

Let us recall that if f is the density function of a non-negative r.v., then the function $f^{(n)}$ is the density function of the sum of n independent r.v. $X_1,...,X_n$ each of them also having f as density function.

ii) It is also possible to use definition [2.2] convolution product in the homogeneous case:

$$\forall 0\le s\le t \text{ and } h:-s<h:$$
$$F(s,t)=F(s+h,t+h),$$
$$G(s,t)=G(s+h,t+h).$$

Given this hypothesis, it is possible to introduce two new functions but now of one variable called $\hat{F},\hat{G}$ and is defined by:

$$\begin{aligned}&\forall s,t,s<t \text{ and } \tau=t-s:\\&\hat{F}(\tau)=F(s,t),\\&\hat{G}(\tau)=G(s,t),\end{aligned}$$

and now

$$(F\otimes G)(s,t)=\int_s^t F(s,\tau)dG(\tau,t)=\int_0^t \hat{F}(t-x)d\hat{G}(x)=(\hat{F}\otimes\hat{G})(t),t\in\mathbb{R}^+. \qquad [2.7]$$

Of course, if F and G have, respectively, functions f and g as densities, clearly, the new d.f. $F\otimes G$ also has a density given by:

$$(F\otimes G)'(t)=(f*g)(t),t\in\mathbb{R}^+. \qquad [2.8]$$

If $F = G$, by the associativity property, we can define successive convolutions products of a non-negative functions f as follows:

$$F^{(2)}(t)=\int_0^t F(t-x)dF(x),$$

$$\vdots$$

$$F^{(n)}(t)=\int_0^t F^{(n-1}(t-x)dF(x).$$

Here too, we have commutativity for F and G distribution functions.

2.3. Homogeneous and non-homogeneous renewal processes

In this section, the main definitions of renewal theory, as given in [JAN 06], are generalized in a non-homogeneous environment. Let $(X_n,\ n\geq 1)$ be a sequence of positive, independent and not necessarily independent or identically distributed r.v. defined on the probability space $(\Omega,\Im,P)$.

DEFINITION 2.3.– *The random sequence* $(T_n, n \geq 0)$, *where:*

$$T_0 = 0 \text{, and } T_n = X_1 + \cdots + X_n, \qquad n \geq 1, \tag{2.9}$$

is called a *generalized renewal sequence* or sometimes *general renewal process*.

The r.v. $T_n, n \geq 0$ are called *renewal times* and the r.v. $X_n, n \geq 1$ are independent and are called *inter-arrival times*. To avoid triviality, it is supposed that $\mathrm{P}[X_n \leq 0 = 0]$.

DEFINITION 2.4.– *The random sequence* $(T_n, n \geq 0)$ *is a homogeneous classical renewal process if the* $X_n, n \geq 1$ *r.v. are identically distributed and independent with* F *as d.f..*

DEFINITION 2.5.– *i) The random sequence* $(T_n, n \geq 0)$ *is a non-homogeneous classical renewal process if variables* $X_n, n \geq 1$ *are conditionally dependent given:*

$$\begin{aligned} &(T_n, n \geq 1) \text{ where} \\ &\mathrm{T}_n = \sum_{i=1}^{n} X_i, n \geq 1, \\ &T_0 = 0 \end{aligned} \tag{2.10}$$

in the following way

$$P(X_n \leq x | T_k, k \leq n-1) = P(X_n \leq x | T_{n-1}), n \geq 1$$

with here the function F is the conditionally d.f. of X_n given T_{n-1} defined as follows:

$$F(s,t) = P(X_n \leq t - s | T_{n-1} = s), 0 < s \leq t \tag{2.11}$$

And for s = 0, we have:

$$F(0,t) = P(X_0 \leq t), 0 \leq t. \tag{2.12}$$

It follows that the process $(T_n, n \geq 0)$ is a non-homogeneous Markov process with values in $\mathbb{R}^+$.

From relation [2.11], we have setting $u = t - s$, we have for *homogeneous classical renewal process* $(T_n, n \geq 0)$:

$$P(T_n \leq t | T_{n-1} = s) = P(X_n \leq t - s | T_{n-1} = s) = F(t - s) \quad [2.13]$$

and we obtain the equidistribution of the X_n

$$P(X_n \leq u | T_{n-1} = s) = F(u + s - u) = F(u).$$

It follows that the d.f. F completely defines the homogeneous renewal process.

REMARK 2.4.– As in the sequel, we will only consider classical homogeneous and non-homogeneous renewal processes, and from now on, we will omit the term "classical".

REMARK 2.5.– For a non-homogeneous renewal process, we successively have:

$$P(X_1 \leq x) = F(0.x)$$

$$P(X_1 \leq x, X_2 \leq y) = \int_0^x P(T_2 - T_1 \leq y | T_1 = u) dF(0,u) \quad [2.14]$$

$$= \int_0^x F(u, u + y) dF(0,u).$$

On the other part:

$$\begin{aligned} P(X_2 \leq u) &= E\left[P(T_2 \leq u + T_1 | T_1 = y)\right] \\ &= \int_0^\infty F(y, u + y) dF(0, y). \end{aligned} \quad [2.15]$$

Combining with relations [2.14], this latter relation proves that, for a non-homogeneous renewal process, the random variables X_1, X_2 are not independent.

Of course in the homogeneous renewal process, we know that:

$F(s,t) = F(t - s)$ and therefore relation [2.14] becomes:

$$P(X_1 \leq x, X_2 \leq y) = \int_0^x F(y) dF(u) = F(y) F(x)$$

And thus, the two r.v are independent. With a similar reasoning, it can be proven that the sequence (X_n) is a sequence of i.i.d. random variables.

DEFINITION 2.6.– *In a delayed homogeneous renewal process,* X_1 *has a different distribution* F_{X_1} *with respect to* F*, the common distribution function of* $X_n, n \geq 2$.

If the sequence $(X_n, n \geq 1)$ of independent r.v. is such that:

$$\begin{aligned} &P(X_i \leq x) = F_i(x), x \geq 0, i = 1,...,k, \\ &P(X_i \leq x) = F(x), x \geq 0, i > k, \end{aligned} \tag{2.16}$$

The process is called a delayed renewal process of order k.

DEFINITION 2.7.– *Let* $\{(X_n, Y_n), n \geq 1\}$ *be a sequence of two-dimensional non-negative r.v. Then, the sequence* $(X_1, Y_1, X_2, Y_2, ..., Y_{n-1}, X_n, Y_n, X_{n+1}, ...)$ *is called a general alternating renewal process. They were introduced by Smith [SMI 58].*

It is called a *classical alternating renewal process* or simply an *alternating renewal process* if $(X_n), (Y_n)$ are two independent sequences of i.i.d. random variables, respectively, with *F* and *G* as distribution function.

2.4. Counting processes and renewal functions

DEFINITION 2.8.– *With each renewal sequence, we can associate the following stochastic process, continuous in time, with values in* $\mathbb{N}$:

$$\left(N(t), \quad t \geq 0\right) \tag{2.17}$$

where

$$N(t) > n-1 \Leftrightarrow T_n \leq t, \qquad n \in \mathbb{N}_0.$$

For every fixed *t*, $N(t)$ represents the total number of "renewals" on $(0,t]$.

These processes are, respectively, called the *homogeneous and non-homogeneous associated counting process* or the *homogeneous and non-homogeneous renewal counting process*.

Writing successively the homogeneous and non-homogeneous case, we clearly have:

$$\begin{aligned} &N(t) > n-1 \Leftrightarrow T_n \leq t, \qquad n \in \mathbb{N} \\ &N(t) - N(s) > n-1 \Leftrightarrow T_{n'} = s, T_{n''} \leq t, \ n'' - n' > n-1. \end{aligned} \tag{2.18}$$

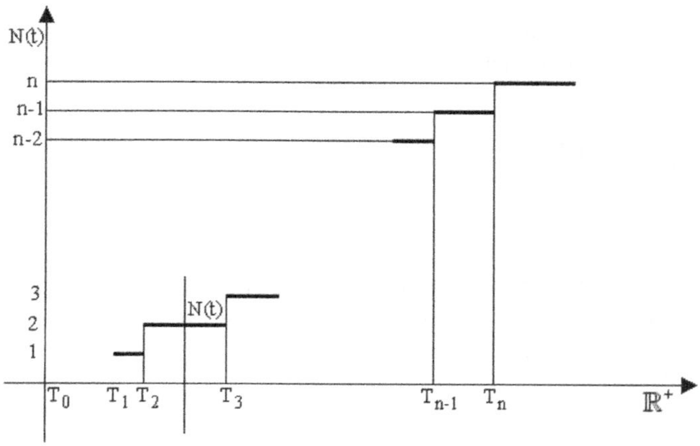

Figure 2.1. *Sample path of a counting process*

REMARK 2.6.– The probability of having at least n renewals within a time t in the homogeneous case and from time s to time t in the non-homogeneous environment is, respectively, given by:

$$\begin{aligned} &\mathrm{P}[N(t) > n-1] = F^{(n)}(t) \\ &\mathrm{P}[N(t) - N(s) > n-1] = F^{(n)}(s,t). \end{aligned} \qquad [2.19]$$

The probability of having just n renewals is obtained in the following way:

$$\mathrm{P}[N(t) = n] = \mathrm{P}[N(t) > n-1] - \mathrm{P}[N(t) > n] = F^{(n)}(t) - F^{(n+1)}(t),$$

$$\mathrm{P}[N(t) - N(s) = n] = \mathrm{P}[N(t) - N(s) > n-1] - \mathrm{P}[N(t) - N(s) > n]$$

and so

$$P[N(t) - N(s) = n] = F^{(n)}(s,t) - F^{(n+1)}(s,t), n \geq 0 \qquad [2.20]$$

with

$$F^{(0)}(s,t) = U_0(t).$$

DEFINITION 2.9.– *The homogeneous and non-homogeneous renewal functions are, respectively, defined as follows:*

$$\begin{aligned} &H(t) = \mathrm{E}[N(t)],\ 0 \leq t,\ t \in \mathbb{R}_0^+, \\ &H(s,t) = E[N(t) - N(s)] = \mathrm{E}[N(t)] - \mathrm{E}[N(s)],\ 0 \leq s \leq t\ \ s,t \in \mathbb{R}_0^+, \end{aligned} \qquad [2.21]$$

provided that the expectation is finite. They give the mean number of renewals that verified, respectively, within a time t and from time s to time t.

$$E\left(\prod_{k=1}^{n} X_k\right) = \prod_{k=1}^{n} E(X_k)$$

PROPOSITION 2.2.– The homogeneous and non-homogeneous continuous time evolution equation of the renewal equations, supposing the absolute continuity of the d.f. F, i.e. $dF(t) = f(t)dt$, can be written, respectively, in the following way:

$$\begin{aligned} H(t) &= F(t) + \int_0^t f(\tau)H(t-\tau)d\tau, \\ H(s,t) &= F(s,t) + \int_s^t f(s,\tau)H(\tau,t)d\tau. \end{aligned} \qquad [2.22]$$

PROOF.– For the homogeneous case see, for example, [JAN 06].

In the non-homogeneous case, it results:

$$\mathrm{E}\left[N(t) - N(s)\right] = H(s,t) = \sum_{n=1}^{\infty} n\left[F^{(n)}(s,t) - F^{(n+1)}(s,t)\right]$$

and therefore

$$H(s,t) = \sum_{n=1}^{\infty} F^{(n)}(s,t) \qquad [2.23]$$

It follows that:

$$\begin{aligned} H(s,t) &= \sum_{n=1}^{\infty} F^{(n)}(s,t) = F(s,t) + \sum_{n=2}^{\infty} F^{(n)}(s,t) \\ &= F(s,t) + \left(\sum_{n=2}^{\infty} F^{(n-1)}\right) * F(s,t) = F(s,t) + \left(\sum_{n=1}^{\infty} F^{(n)}\right) * F(s,t) \\ H(s,t) &= F(s,t) + H * F(s,t) \end{aligned}$$

$$H(s,t) = F(s,t) + \int_s^t f(s,\tau)H(\tau,t)d\tau$$ □

REMARK 2.7.–

1) It is possible to show (see [JAN 06]) that:

$$\mathrm{Var}(N(t)) = H(t) + 2H^{(2)}(t) - (H(t))^2 \quad [2.24]$$

2) Relation [2.23] gives an explicit form of the renewal function of a general renewal function under the form of a series but of convolution products and therefore not easy to compute, and this is why it is better to numerically solve the integral renewal equation [2.23]. Particular cases of relation [2.23] are:

$$H(t) = \sum_{n=1}^{\infty} F^{(n)}(t) \text{ for a renewal process,}$$

$$\begin{aligned} &H_1(t) = F_1 + \sum_{n=1}^{\infty} F_1 \otimes F^{(n)}(t) \text{ for a delayed renewal process,} \\ &H_k(t) = F_1 + F_1 \otimes F_2 + \cdots + F_1 \otimes F_2 \otimes \cdots \otimes F_k(t) \\ &\qquad + F_1 \otimes F_2 \otimes \cdots \otimes F_k \otimes F + F_1 \otimes F_2 \otimes \cdots \otimes F_k \otimes \sum_{n=1}^{\infty} F^{(n)}, \end{aligned} \quad [2.25]$$

for a delayed renewal process of order K.

PROPOSITION 2.3.– For a delayed homogeneous renewal process for which the d.f. of X_1 is F_1 and the one of the r.v. $X_n, n \geq 2$ is F, the renewal function H_1 of this process is related to the renewal function of the renewal process H defined by F with the following integral relation:

$$H_1(t) = F_1(t) + \int_0^t H(t-u)dF_1(u). \quad [2.26]$$

More generally, for a delayed process of order k, we have with the notations of definition 2.6

$$H_k(t) = F_1(t) + F_1 \otimes F_2(t) + \ldots + F_1 \otimes F_2(t) \otimes \ldots \oplus F_k(t) + F_i \otimes H(t) \quad [2.27]$$

where the functions $F_1, \ldots, F_k$ are given by relation [2.17] and H the renewal function of the homogeneous renewal function characterizes by F.

PROOF.– For a general homogeneous renewal process for which each r.v has a different d.f:

$$P(X_i \leq x) = F_i(x), i \geq 1,$$

we have

$$\begin{aligned} E[N(t)] &= \sum_{j=1}^{\infty} j[F_1 \otimes ... \otimes F_i)(t) - F_1 \otimes ... \otimes F_{i+1})(t)] \\ &= \sum_{j=1}^{\infty} F_1 \otimes ... \otimes F_j)(t) \end{aligned} \quad [2.28]$$

And thus for a delayed renewal process, we have:

$$\begin{aligned} E[N(t)] &= \sum_{j=1}^{\infty} F_1 \otimes F \otimes ... F \otimes F)(t) \\ &= F_1(t) + F_1 \otimes H(t), \end{aligned} \quad [2.29]$$

This is relation [2.26].

Now, a delayed process of order k, we have:

$$\begin{aligned} &P(Y_i \leq x) = (F_1 \otimes ... \otimes F_i)(t), i \leq k, \\ &P(Y_j \leq x) = F(x), j > k. \end{aligned} \quad [2.30]$$

As from relation [2.25] for a homogeneous process:

$$\begin{aligned} &H_k(t) = F_1(t) + F_1 \otimes F_2(t) + \cdots + F_1 \otimes F_2 \cdots \otimes F_k \\ &+ F_1 \otimes F_2 \cdots \otimes F_k \otimes F + F_1 \otimes F_2 \cdots \otimes F_k \otimes F^{(2)} + \cdots, \end{aligned}$$

we obtain

$$H_k(t) = F_1(t) + F_1 \otimes F_2(t) + \cdots + F_1 \otimes F_2 \cdots \otimes F_k \otimes H(t)$$

This is relation [2.27]. □

For numerical application, it is often necessary to consider stochastic processes in discrete time, that is for the set of possible values of the time parameter t, for example $\mathbb{N}$.

DEFINITION 2.10.– *The homogeneous and non-homogeneous discrete time evolution equations of the renewal processes are, respectively, given by:*

$$\begin{aligned} H(t) &= F(t) + \sum_{x=1}^{t} \nu(\tau) H(t-\tau) \\ H(s,t) &= F(s,t) + \sum_{x=s+1}^{t} \nu(s,\tau) H(\tau,t) \end{aligned} \qquad [2.31]$$

where $F(t)$ and $F(s,t)$ are, respectively, the waiting time for the homogeneous and non-homogeneous d.f. They give the probability of having a renewal within the time t, given that the system was followed from time 0 in the homogeneous case and from time s in the non-homogeneous environment. Furthermore, it results that:

$$\begin{aligned} \nu(t) &= \begin{cases} 0 & \text{if } t = 0 \\ F(t) - F(t-1) & \text{if } t > 0, \end{cases} \\ \nu(s,t) &= \begin{cases} 0 & \text{if } s \geq t \\ F(s,t) - F(s,t-1) & \text{if } s < t. \end{cases} \end{aligned} \qquad [2.32]$$

REMARK 2.8.– In the homogeneous case, the definitions of continuous and discrete time evolution equations correspond to those given, respectively, in [FEL 71] (p. 185) and [FEL 68] (p. 332).

2.5. Asymptotical results in the homogeneous case

Since the publication of the seminal papers of Smith [SMI 54, SMI 58], the asymptotic behavior of a renewal theory is well known. We will only give the fundamental results whose detailed proofs can be found in [JAN 06].

PROPOSITION 2.4.– The basic asymptotical results are:

1) For a homogeneous renewal theorem characterized by the d.f. F with mean m, we have:

$$\lim_{t \to \infty} \frac{N(t)}{t} = \frac{1}{m}, \ \textit{almost surely}$$

2) For a homogeneous renewal theorem characterized by the d.f. F with mean m, we have:

$$\lim_{t \to \infty} \frac{H(t)}{t} = \frac{1}{m}. \qquad [2.33]$$

3) Blackwell's theorem: for a homogeneous renewal theorem characterized by the d.f. F with mean m, we have, for every positive number τ :

$$\lim_{t\to\infty}\left(R(t)-R(t-\tau)\right)=\frac{\tau}{m}. \tag{2.34}$$

4) For a homogeneous renewal theorem characterized by the d.f. F with mean m and variance σ^2, we have:

$$\lim_{t\to\infty}\left(H(t)-\frac{t}{m}\right)=\frac{\sigma^2-m^2}{2m^2}. \tag{2.35}$$

or

$$H(t)=\frac{t}{m}+\frac{\sigma^2-m^2}{2m^2}+O(1), \tag{2.36}$$

where O(1) represents a function of t approaching zero as t approaches infinity.

PROPOSITION 2.5.– (Central limit theorem).

If $\sigma^2<\infty$, *then for all* $y\in\mathbb{R}$:

$$\lim_{t\to\infty}P\left(\frac{N(t)-t/m}{\sqrt{\sigma^2\,t/m^3}}\le y\right)=\Phi(y)\ , \tag{2.37}$$

where Φ *represents the standard normal d.f.*

REMARK 2.9.– i) From proposition 2.5, we have the following approximation for large t :

$$\mathrm{var}(N(t))\sim\frac{\sigma^2}{m^3}t, \tag{2.38}$$

which simplifies the exact result [2.24].

ii) The main result used to obtain these asymptotical results is the key renewal theorem saying that:

PROPOSITION 2.6.– (Asymptotical behavior and key renewal theorem)

i) *In the transient case, we have*:

$$\lim_{t\to\infty} X(t) = R(\infty)G(\infty) \tag{2.39}$$

provided that the limit

$$G(\infty) = \lim_{t} G(t)$$

exists.

ii) *In the case of recurrence, we have*:

$$\lim_{t\to\infty} X(t) = \frac{1}{m}\int_0^\infty G(x)dx, \tag{2.40}$$

provided that G is directly Riemann integrable on $[0,\infty)$, *that F is not arithmetic, and supposing*:

$$m = E(X_n) = \int_0^\infty (1-F(x))\,dx. \tag{2.41}$$

COROLLARY 2.1.– In the case of a recurrent renewal process with finite variance σ^2, we have:

$$\lim_{t\to\infty}\left(R(t) - \frac{t}{m}\right) = \frac{m^2+\sigma^2}{2m^2}. \tag{2.42}$$

2.6. Recurrence times in the homogeneous case

Let us consider the renewal process $(T_n,\ n \geq 0)$ characterized by the d.f. F of mean m.

At a positive time t, we know that the lifetime is that of $X_{N(t)+1}$ or $X_{N(t)}$, as:

$$S_{N(t)} \leq t \leq S_{N(t)+1}.$$

The r.v. $\delta(t)$ called *backward recurrence time* or *age* at time t is defined as:

$$\delta(t) = t - S_{N(t)}, \tag{2.43}$$

and the time to wait for the next renewal, say $\gamma(t)$, is called the *excess* or *forward recurrence time*, or *excess* or sometimes the *residual life* at time t:

$$\gamma(t) = S_{N(t)+1} - t. \qquad [2.44]$$

Of course, the total time spent between the two successive renewals, called the *recurrence time* or *spread* $X_{N(t)+1}$, satisfies:

$$X_{N(t)+1} = \delta(t) + \gamma(t). \qquad [2.45]$$

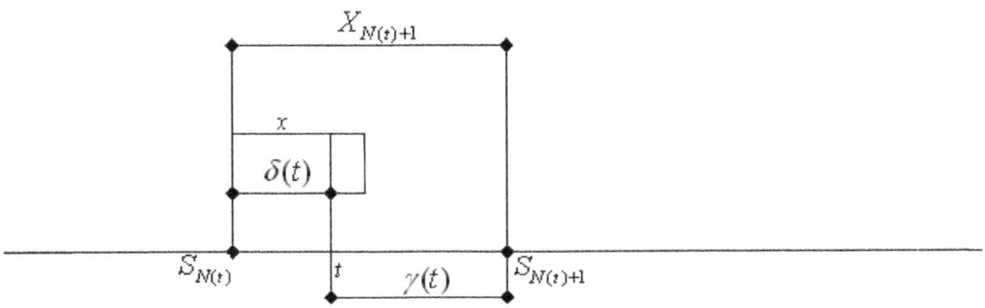

Figure 2.2. *Backward (age) and forward (excess) recurrence times and spread*

In the homogeneous case, the following results related to the distributions of r.v. $\delta(t)$, $\gamma(t)$ and $X_{N(t)+1}$ for all t may be found in [JAN 06].

PROPOSITION 2.7.– Distribution forward recurrence time

If $F_{\delta(t)}$ is the d.f. of age $\delta(t)$, we have:

$$F_{\delta(t)}(x) = \begin{cases} \int_{[t-x,t]} [1 - F(t-u)]\, dH(u), x \le t, \\ 1, x > t. \end{cases} \qquad [2.46]$$

Distribution of backward recurrence time

$$\mathrm{P}(\gamma(t) \le x) = F(t+x) - F(t) + \int_{[0,t]} [F(t-u+x) - F(t-u)]\, dH(u) \qquad [2.47]$$

Distribution of the spread

$$F_{X_{N(t)+1}}(x)=\begin{cases}\int_{t-x}^{t}[F(x)-F(t-u)]dH(u), & \text{if } x\le t,\\ F(x)-F(t)+\int_{0}^{t}[F(x)-F(t-u)]dH(u), & \text{if } x>t.\end{cases} \qquad [2.48]$$

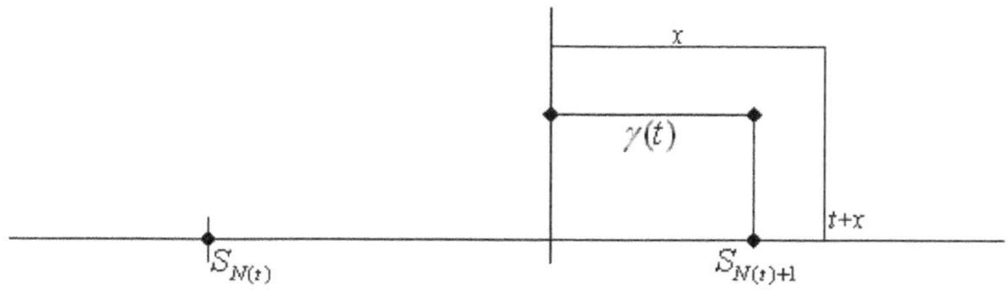

Figure 2.3. *Age events analysis*

If we examine Figure 2.3, we can deduce that event $\gamma(t)\le x$ is realized iff:

$$S_{N(t)}\le t<S_{N(t)+1}\le t+x.$$

REMARK 2.10.– The expectation of $\gamma(t)$ is given by:

$$E(\gamma(t))=m[H(t)+1]-t. \qquad [2.49]$$

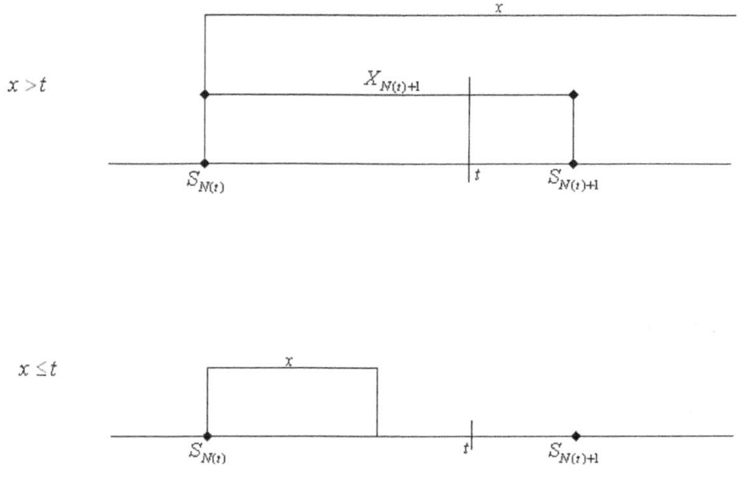

Figure 2.4. *Spread events analysis*

Suppose that for all t, the recurrence times $\gamma(t)$ and $\delta(t)$ are independent; then the renewal process is a Poisson process.

2.7. Particular case: the Poisson process

2.7.1. *Homogeneous case*

This process was already introduced in Chapter 1. Clearly, a Poisson process is a homogeneous renewal process defined by the d.f. G defined by:

$$G(t) = \begin{cases} 1 - e^{-\lambda t}, t \geq 0, \\ 0, \qquad t < 0, \end{cases} \qquad [2.50]$$

where λ is a parameter strictly positive and constant.

Using results [2.20] and [2.23], we have:

$$P[N(t) = k] = \frac{(\lambda t)^k}{k!} e^{-\lambda t}, k = 0,1,... \qquad [2.51]$$

and

$$H(t) = E[N(t)] = \lambda t. \qquad [2.52]$$

This renewal function is the unique solution of the renewal equation [2.22]. As any renewal function characterizes a renewal process, the Poisson process is characterized by a linear renewal function.

Using proposition 2.7, we obtain:

$$F_{\delta(t)}(x) = \begin{cases} \lambda \int\limits_{[t-x,t)} e^{-\lambda(t-u)} du, & x < t, \\ 1 & , \ x \geq t \ . \end{cases} \qquad [2.53]$$

After computation of the elementary definite integral, we obtain:

$$F_{\delta(t)}(x) = \begin{cases} 1 - e^{-\lambda x}, & x < t, \\ 1 & , \ x \geq t \ . \end{cases} \qquad [2.54]$$

For the distribution of the excess $\gamma(t)$, we have:

$$F_{\gamma(t)}(x)=\lambda\int_0^t\left[e^{-\lambda(t-u)}-e^{-\lambda(t-u+x)}\right]du+e^{-\lambda t}-e^{-\lambda(t+x)}$$

$$=\lambda e^{-\lambda t}\int_0^t e^{\lambda u}\left(1-e^{-\lambda x}\right)du+e^{-\lambda t}\left(1-e^{-\lambda x}\right)$$

$$=\lambda e^{-\lambda t}\left(1-e^{-\lambda x}\right)\int_0^t e^{\lambda u}du+e^{-\lambda t}\left(1-e^{-\lambda x}\right)$$

$$=e^{-\lambda t}\left(1-e^{-\lambda x}\right)\left(e^{\lambda t}-1\right)+e^{-\lambda t}\left(1-e^{-\lambda x}\right)$$

$$=\left(1-e^{-\lambda x}\right)\left(1-e^{-\lambda t}\right)+e^{-\lambda t}\left(1-e^{-\lambda x}\right),$$

and finally

$$F_{\gamma(t)}(x)=\left(1-e^{-\lambda x}\right)=F(x). \qquad [2.55]$$

The latter result proves that the distributions of the age and the excess are independent of t and that it is the original d.f., which is in fact equivalent to the memoryless property of the exponential distribution.

For the d.f. of the spread, we have from proposition 2.7 that:

$$F_{X_{N(t)+1}}(x)=\begin{cases}\lambda\int_{t-x}^t\left[e^{-\lambda(t-u)}-e^{-\lambda x}\right]du, & \text{if } x\le t,\\ e^{-\lambda t}-e^{-\lambda x}+\lambda\int_0^t\left[e^{-\lambda(t-u)}-e^{-\lambda x}\right]du, & \text{if } x>t.\end{cases}$$

As

$$\lambda\int_{t-x}^t\left[e^{-\lambda(t-u)}-e^{-\lambda x}\right]du=\lambda\left[\frac{1-e^{-\lambda x}}{\lambda}-e^{-\lambda x}x\right]$$

$$=1-e^{-\lambda x}-\lambda xe^{-\lambda x}$$

$$=1-(1+\lambda x)e^{-\lambda x}$$

and

$$\lambda\int_0^t\left[e^{-\lambda(t-u)}-e^{-\lambda x}\right]du=\lambda\left[\frac{1-e^{-\lambda t}}{\lambda}-e^{-\lambda x}t\right],$$

we finally get that:

$$F_{X_{N(t)+1}}(x)=\begin{cases}1-(1+\lambda x)e^{-\lambda x}, & \text{if } x\leq t,\\ 1-\lambda^2te^{-\lambda x}-e^{-\lambda x}, & \text{if } x>t,\end{cases}\qquad [2.56]$$

$$f_{X_{N(t)+1}}(x)=\begin{cases}\lambda^2xe^{-\lambda x}, & \text{if } x\leq t,\\ \lambda^3te^{-\lambda x}+\lambda e^{-\lambda x}, & \text{if } x>t.\end{cases}\qquad [2.57]$$

2.7.2. *Non-homogeneous case*

If in the assumption given in section 2.7.1, we replace the parameter λ by a non-negative measurable function of t, say $\lambda(t)$, the new defined process is called a *non-homogeneous Poisson process of parameter* $\lambda(t)$.

In fact, we obtain a non-homogeneous renewal process characterized by the function $F(.,.)$ given by relation [2.11], thus with:

$$F(s,t)=P(T_n\leq t|T_{n-1}=s)(=P(X_n\leq t-s|T_{n-1}=s)),$$

F being defined by:

$$F(s,t)=1-e^{-\int_s^t\lambda((u)du}, s\leq t.\qquad [2.58]$$

From result [2.20], it follows that for each fixed t, $N(t)$ has a Poisson distribution of parameter $\int_0^t\lambda(u)du$ and for fixed s and t, $s<t, N(t)-(s)$ a Poisson distribution of parameter $\int_s^t\lambda(u)du$.

REMARK 2.11.– As we successively have:

$$P(X_1 \le x) = F(0,x) = 1 - e^{-\int_0^x \lambda((z)dz},$$

$$P(X_2 \le x) = \int_0^\infty P(T_2 - T_1 \le x | T_1 = s) dF(0,s)$$

$$= \int_0^\infty (1 - e^{-\int_s^{s+x} \lambda((z)dz}) de^{-\int_0^s \lambda((z)dz}$$

$$= \int_0^\infty (1 - e^{-\int_s^{s+x} \lambda((z)dz}) e^{-\int_0^s \lambda((z)dz} \lambda(s)ds$$

$$= 1 - \int_0^\infty e^{-\int_s^{s+x} \lambda((z)dz} e^{-\int_0^s \lambda((z)dz} \lambda(s)ds$$

$$= 1 - \int_0^\infty e^{-\int_0^{s+x} \lambda((z)dz} \lambda(s)ds$$

$$= 1 - e^{-\int_0^x \lambda((z)dz} \int_0^\infty e^{-\int_x^{s+x} \lambda((z)dz} \lambda(s)ds$$

and as $\int_0^\infty e^{-\int_x^{s+x} \lambda((z)dz} \lambda(s)ds = 1$

we finally have:

$$= P(X_1 \le x).$$

It follows that the r.v. X_1, X_2 are identically distributed.

Similarly, we can prove that the r.v. $(X_n, n \ge 1)$ are identically distributed. However, starting from relation [2.15], we can also show that X_1, X_2 are independent. By induction, we can finally prove that the r.v. are independent and identically distributed though we have a non-homogeneous renewal process.

2.8. Homogeneous alternating renewal processes

Up to now, it was assumed that renewals start as soon as they happen. In the real world, it is possible that this condition is not always satisfied, that is renewals can start after a non-negligible random time. It is possible to take into account this phenomenon defining the concept of alternating renewal process. Indeed, this process mixes two renewal processes in the following sense. Let us consider two random sequences $\{Y_1,Y_2,\ldots\}$ and $\{Z_1,Z_2,\ldots\}$. In this way, the sequence of two-dimensional random vectors $\{(Y_1,Z_1),(Y_2,Z_2),\ldots,(Y_k,Z_k),\ldots\}$ $Y_i,Z_i\in\mathbb{N}$ is defined to be a *general alternating renewal process*, for which the times of renewals are given by the sequence $\{S_1,T_1,S_2,S_3,\ldots,S_n,T_n,\ldots\}$, where

$$S_1=Y_1,\ S_n=\sum_{i=1}^{n-1}(Y_i+Z_i)+Y_n,n=2,3,\ldots, \qquad [2.59]$$

$$T_n=\sum_{i=1}^{n}(Y_i+Z_i),n=1,2,\ldots;\ T_n\in\mathbb{N} \qquad [2.60]$$

They successively represent the alternating renewal times of the two sequences $\{Y_1,Y_2,\ldots\}$ and $\{Z_1,Z_2,\ldots\}$.

When the two sequences $\{Y_1,Y_2,\ldots\}$ and $\{Z_1,Z_2,\ldots\}$ are independent and formed both of i.i.d. random variables with $F_Y(y)=\mathrm{P}(Y\le y)$ and $F_Z(z)=\mathrm{P}(Z\le z)$ respectively as distribution functions, they form what we called a *classical alternating renewal process* or simply an *alternating renewal process* thus defined by the two d.f F_X,F_Y .

For a fixed t, let $N_S(t)$ and $N_T(t)$ represent the number of successive renewals that happened in $(0,t]$, respectively, for the S- and T-process. From the independence assumptions, it follows that S_n and T_n are the sums of independent r.v. and thus:

$$F_{S_n}(t)=\mathrm{P}(S_n\le t)=\mathrm{P}(N_S(t)\ge n)=F_Y*(F_Y*F_Z)^{(n-1)}(t) \qquad [2.61]$$

$$F_{T_n}(t)=\mathrm{P}(T_n\le t)=\mathrm{P}(N_T(t)\ge n)=(F_Y*F_Z)^{(n)}(t).$$

Consequently, the two renewal functions defined by:

$$H_S(t)=\mathrm{E}(N_S(t))\text{ and }H_T(t)=\mathrm{E}(N_T(t))$$

are given by

$$\begin{aligned} H_S(t) &= \sum_{n=1}^{\infty} F_Y * (F_Y * F_Z)^{(n-1)}(t) = F_Y(t) + \sum_{n=2}^{\infty} F_Y * (F_Y * F_Z)^{(n-1)}(t) \\ &= F_Y(t) + \sum_{n=1}^{\infty} F_Y * (F_Y * F_Z)^{(n)}(t) \end{aligned} \quad [2.62]$$

$$H_T(t) = \sum_{n=1}^{\infty} (F_Y * F_Z)^{(n)}(t). \quad [2.63]$$

$H_S(t)$ can be seen as a renewal function of the delayed renewal process in which the first renewal is distributed as F_Y and the other renewals are distributed as F_{Y+Z}.

In the same way, $H_r(t)$ can be seen as the renewal function of the renewal process characterized by the d.f. F_{Y+Z}.

If we are interested in the total number of renewals $N(t)$, we clearly have $N(t) = N_S(t) + N_T(t)$ and therefore:

$$E[N(t)] = H_S(t) + H_T(t) \quad [2.64]$$

2.9. Solution of non-homogeneous discrete time evolution equation

2.9.1. *General method*

This section presents a straightforward method that deals directly with the numerical solution of the renewal equation. It proves that any kind of renewal process can be solved by means of this approach and that, in a very particular case of the general numerical solution (general because the d.f. is not specified), the related discrete time renewal equation can be obtained.

Freiberger and Grenander [FRE 71] presented this approach in the homogeneous case without any theoretical justification, as did Xie [XIE 89] in which the so-called "midpoint" formula was given with neither a general introduction nor a theoretical justification. Many other papers deal with the problem of discretization of the homogeneous renewal equation, see for example [DEV 96] and [ELK 01], but as far as the authors are aware none of these papers state that it is possible to obtain the discrete time renewal equation from the continuous one and the continuous from the discrete environment. More recently, Moura and Droguett [MOU 08] and Droguett *et al.* [DRO 08] and Droguett and Moura [DRO 09] presented a faster algorithm, which simplifies the problem and is useful for solving the non-homogeneous and homogeneous semi-Markov evolution equation developed in Chapter 4. The results are a direct function of the observed data.

The discrete time non-homogeneous renewal equations [2.22] can be numerically solved very easily (for the homogeneous case, see [JAN 06]). Using matrix notations, the non-homogeneous case of relation [2.22] can be written as:

$$H(s,t)-\sum_{x=s+1}^{t} \nu(s,\tau)H(\tau,t)=F(s,t) \Leftrightarrow \mathbf{U}\cdot\mathbf{H}=\mathbf{F} \qquad [2.65]$$

With the following matrices:

$$\mathbf{U}=\begin{bmatrix} 1 & -\nu_{0,1} & \cdots & -\nu_{0,k} & \cdots & -\nu_{0,n} & \cdots \\ 0 & 1 & \cdots & -\nu_{1,k} & \cdots & -\nu_{1,n} & \cdots \\ \vdots & \vdots & \ddots & \vdots & \ddots & \vdots & \ddots \\ 0 & 0 & \cdots & 1 & \cdots & -\nu_{k,n} & \cdots \\ \vdots & \vdots & \ddots & \vdots & \ddots & \vdots & \ddots \\ 0 & 0 & \cdots & 0 & \cdots & 1 & \cdots \\ \vdots & \vdots & \ddots & \vdots & \ddots & \vdots & \ddots \end{bmatrix}, \qquad [2.66]$$

$$\mathbf{H}=\begin{bmatrix} H_{0,0} & H_{0,1} & \cdots & H_{0,k} & \cdots & H_{0,n} & \cdots \\ 0 & H_{1,1} & \cdots & H_{1,k} & \cdots & H_{1,n} & \cdots \\ \vdots & \vdots & \ddots & \vdots & \ddots & \vdots & \ddots \\ 0 & 0 & \cdots & H_{k,k} & \cdots & H_{k,n} & \cdots \\ \vdots & \vdots & \ddots & \vdots & \ddots & \vdots & \ddots \\ 0 & 0 & \cdots & 0 & \cdots & H_{n,n} & \cdots \\ \vdots & \vdots & \ddots & \vdots & \ddots & \vdots & \ddots \end{bmatrix},$$

$$\mathbf{F}=\begin{bmatrix} F_{0,0} & F_{0,1} & \cdots & F_{0,k} & \cdots & F_{0,n} & \cdots \\ 0 & F_{1,1} & \cdots & F_{1,k} & \cdots & F_{1,n} & \cdots \\ \vdots & \vdots & \ddots & \vdots & \ddots & \vdots & \ddots \\ 0 & 0 & \cdots & F_{k,k} & \cdots & F_{k,n} & \cdots \\ \vdots & \vdots & \ddots & \vdots & \ddots & \vdots & \ddots \\ 0 & 0 & \cdots & 0 & \cdots & F_{n,n} & \cdots \\ \vdots & \vdots & \ddots & \vdots & \ddots & \vdots & \ddots \end{bmatrix},$$

Thus, **U** is a coefficient matrix, **H** is the unknown matrix of renewal functions and **F** is the matrix in which each row is a discrete time d.f.

From a result of [RIE 13], the linear system [2.65] has a unique solution as the determinant of the coefficient matrix of the system [2.65] is equal to 1.

Let us point out that $F(k,k)=0$ because it is impossible to have a renewal in a time 0. This implies $v(k,k)=0$ and $H(k,k)=0$.

2.9.2. *Some particular formulas*

In this section, some formulas of the numerical solution of the non-homogeneous part of renewal equations [2.65] will be given. The relations will be related to particular generalized Newton–Cotes formulas (see [HIL 87]).

The Simpson quadrature method gives the following equations:

$$\begin{aligned}\hat{H}(uh,kh) &= F(uh,kh)+\frac{h}{3}\hat{H}(uh,kh)f(uh,uh)\\ &+\frac{4h}{3}\sum_{\tau=\left\lfloor\frac{u}{2}\right\rfloor+1}^{\left\lfloor\frac{k}{2}\right\rfloor}\hat{H}\left((2\tau-1)h,kh\right)f\left(uh,(2\tau-1)h\right)\\ &+\frac{2h}{3}\sum_{\tau=\left\lfloor\frac{u}{2}\right\rfloor+1}^{\left\lceil\frac{k}{2}\right\rceil-1}\hat{H}\left(2\tau h,kh\right)f\left(uh,2\tau h\right)+\frac{h}{3}\hat{H}(kh,kh)f(uh,kh),\end{aligned} \qquad [2.67]$$

where $\hat{H}$ represents the approximate value of H and h represents the discretization interval.

Using Bezout's quadrature method, the following relation is obtained:

$$\begin{aligned}\hat{H}(uh,kh) &= F(uh,kh)+\frac{h}{2}\hat{H}(uh,kh)f(uh,uh)\\ &+h\sum_{\tau=u+1}^{k-1}\hat{H}\left(\tau h,kh\right)f\left(uh,\tau h\right)\\ &+\frac{h}{2}\hat{H}(kh,kh)f(uh,kh).\end{aligned} \qquad [2.68]$$

Finally, if the simplest quadrature method (rectangle formula) is applied, then it is possible to obtain two different formulas; one giving the value of the integrating function at the end of the interval and the other giving the value at the beginning of the interval. In this way, the following formulas are obtained:

$$\hat{H}(uh,kh)=F(uh,kh)+h\sum_{\tau=u+1}^{k}\hat{H}(\tau h,kh)f(uh,\tau h).$$

$$\hat{H}(uh,kh)=F(uh,kh)+h\sum_{\tau=u}^{k-1}\hat{H}(\tau h,kh)f(uh,\tau h).$$

Substituting the differential $(hf(uh,kh))$ by means of the difference, it, respectively, results:

$$\begin{aligned}\hat{H}(uh,kh)&\cong F(uh,kh)+\sum_{\tau=u+1}^{k}\hat{H}(\tau h,kh)\big(F(uh,\tau h)-F(uh,(\tau-1)h)\big),\\ \hat{H}(uh,kh)&\cong F(uh,kh)+\sum_{\tau=u}^{k-1}\hat{H}(\tau h,kh)\big(F(uh,(\tau+1)h)-F(uh,(\tau)h)\big).\end{aligned}\qquad [2.69]$$

REMARK 2.12.– In [BAK 77] (see p. 925), there are two lemmas and a theorem ensuring that, under our conditions, the approximation of [2.67] to the solution tends to 0 as the discretization interval h tends to 0. However, given our particular system, in the following we are able to give a more interesting result.

2.9.3. *Relations between discrete time and continuous time renewal equations*

Posing h = 1 in [2.69]:

$$\hat{H}(u,k)\cong F(u,k)+\sum_{\tau=u+1}^{k}\hat{H}(\tau,k)\big(F(u,\tau)-F(u,\tau-1)\big).\qquad [2.70]$$

Furthermore, if

$$v(u,\tau)=\begin{cases}F(u,\tau)=0 & \tau=u,\\ F(u,\tau)-F(u,\tau-1) & \tau>u,\end{cases}$$

then relation [2.65] becomes:

$$H(u,k) = F(u,k) + \sum_{\tau=1}^{k} H(\tau,k)\, v(u,\tau), \qquad [2.71]$$

this is the discrete time non-homogeneous renewal equation.

Now, let H be a continuous time renewal function and $\{T_n\}$ be the related renewal process.

If we pose:

$$T_n^h = \left\lfloor \frac{T_n}{h} \right\rfloor h \qquad [2.72]$$

and

$$N^h(t) = n \text{ if } T_n^h \leq t < T_{n+1}^h, \qquad [2.73]$$

then the related discrete time renewal function is given by:

$$H^h(uh,kh) = F^h(uh,kh) + \sum_{\tau=u+1}^{k} v^h(uh,\tau h) H^h\left((k-\tau)h,kh\right).$$

The renewal process T_n^h is defined in the same probability space (Ω, F, P) of T_n.

PROPOSITION 2.8.– The T_n^h process converges a.s. to T_n for $h \to 0$. $\boldsymbol{\mu}, \boldsymbol{\Sigma}$

PROOF.– From the definitions given in section 2.3 and [2.39], it results:

$$P\left[\underset{h\to 0}{N^h(t) = n} \right] \to P\left[N(t) = n\right] \quad \forall n \in \mathbb{N}, t \in \mathbb{R}^+. \qquad [2.74]$$

T is the last result which implies:

$$\underset{h\to 0}{T_n^h} \xrightarrow{a.s.} T_n.$$

□

REMARK 2.13.– Proposition 2.8 can be considered a very special case of theorem 2.10 given in [JAN 01]. However, in this case, the proof is trivial.

REMARK 2.14.–

i) The results obtained show that it is possible to obtain the non-homogeneous discrete time renewal equation by means of the simplest discretization of the related continuous time and that, starting from the discrete time, it is also possible to obtain the continuous time.

ii) Freiberger and Grenander [FRE 71] demonstrate a similar approach to the renewal equation numerical solution for the homogeneous case, but there is no justification of the method. Many other papers deal with the same problem in the homogeneous case, but, as far as the authors know, the relationship between the discrete time and continuous time homogeneous renewal process has only been justified in the book by Janssen and Manca [JAN 06]. It is, however, the first time that the numerical treatment of the non-homogeneous renewal processes is presented.

3

Markov Chains

This chapter presents some fundamental results concerning the theory of Markov chains with a finite number of states both for homogeneous and non-homogeneous cases; most of them will be used in Chapter 4, which is devoted to the extension to semi-Markov processes.

3.1. Definitions

Let us consider an economic or physical system S with m possible states, represented by the set I :

$$I = \{1, \ 2, \ldots, \ m\} . \qquad [3.1]$$

Let the system S evolve randomly in discrete time $(t = 0, \ 1, \ 2, \ldots, \ n, \ldots)$, and let J_n be the r.v. representing the state of the system S at time n.

DEFINITION 3.1.– *The random sequence $(J_n, \ n \in \mathbb{N})$ is a Markov chain iff for all $j_0, \ j_1, \ldots, j_n \in I$:*

$$P(J_n = j_n \mid J_0 = j_0, \ J_1 = j_1, \ldots, \ J_{n-1} = j_{n-1}) = P(J_n = j_n \mid J_{n-1} = j_{n-1}) \qquad [3.2]$$

(provided that this probability has meaning).

DEFINITION 3.2.– *A Markov chain $(J_n, \ n \geq 0)$ is homogeneous iff the probabilities [3.2] do not depend on n and is non-homogeneous in the other cases.*

3.2. Homogeneous case

3.2.1. *Basic definitions*

In the homogeneous *case*, we write:

$$P(J_n = j \mid J_{n-1} = i) = p_{ij}, \qquad [3.3]$$

and we introduce the matrix **P** defined as:

$$\mathbf{P} = \left[p_{ij} \right]. \qquad [3.4]$$

The elements of the matrix **P** have the following properties:

i) $p_{ij} \geq 0, \ \text{ for all } i, j \in I,$ [3.5]

ii) $\sum_{j \in I} p_{ij} = 1, \ \text{ for all } i \in I.$ [3.6]

A matrix **P** satisfying these two conditions is called a *Markov matrix* or a *transition matrix*.

With every transition matrix, we can associate a *transition graph* where vertices represent states. There exists an *arc* between vertices *i* and *j* iff $p_{ij} > 0$.

To fully define the evolution of a Markov chain, it is also necessary to fix an *initial distribution* for state J_0, i.e. a vector:

$$\mathbf{p} = (p_1, \ldots, p_m), \qquad [3.7]$$

such that:

$$\begin{aligned} & p_i \geq 0, \qquad i \in I, \\ & \sum_{i \in I} p_i = 1. \end{aligned} \qquad [3.8]$$

For all i, p_i represents the *initial probability* of starting from i:

$$p_i = P(J_0 = i). \qquad [3.9]$$

So, any homogeneous Markov chains can be characterized by the couple $(\mathbf{p}, \mathbf{P})$.

If $J_n = i$ a.s., that is if the system starts with probability equal to 1 from state i, then the components of vector **p** will be $p_j = \delta_{ij}$.

We now introduce the *transition probabilities of order* $p_{ij}^{(n)}$, defined as:

$$p_{ij}^{(n)} = P\left(J_{\nu+n} = j \mid J_\nu = i\right). \qquad [3.10]$$

From the Markov property [3.2], it is clear that conditioning with respect to $J_{\nu+1}$, we get:

$$p_{ij}^{(2)} = \sum_k p_{ik} p_{kj}. \qquad [3.11]$$

Using the following matrix notation:

$$\mathbf{P}^{(2)} = \left[p_{ij}^{(2)}\right], \qquad [3.12]$$

we find that relation [3.11] is equivalent to

$$\mathbf{P}^{(2)} = \mathbf{P}^2. \qquad [3.13]$$

Using induction, it is easy to prove that if:

$$\mathbf{P}^{(n)} = \left[p_{ij}^{(n)}\right], \qquad [3.14]$$

then we obtain for all $n \geq 1$:

$$\mathbf{P}^{(n)} = \mathbf{P}^n. \qquad [3.15]$$

So, the transition probability matrix in n steps is equal to the nth power of the matrix **P**.

For the marginal distributions related to J_n, let us define for $i \in I$ and $n \geq 0$:

$$p_i(n) = P\left(J_n = i\right). \qquad [3.16]$$

These probabilities may be computed as follows:

$$p_i(n) = \sum_j p_j p_{ji}^{(n)}, \qquad i \in I. \qquad [3.17]$$

If we write:

$$p_{ji}^{(0)} = \delta_{ji} \text{ or } \mathbf{P}^{(0)} = \mathbf{I}, \qquad [3.18]$$

then relation [3.15] is true for all $n \geq 0$.

If:

$$\mathbf{p}(n) = \left(p_1(n), \ldots, p_m(n) \right), \qquad [3.19]$$

then relation [3.17] can be expressed, using matrix notation, as:

$$\mathbf{p}(n) = \mathbf{p}\mathbf{P}^n. \qquad [3.20]$$

DEFINITION 3.3.– *A Markov matrix **P** is regular if there exists a positive integer* k, *such that all the elements of the matrix* $\mathbf{P}^{(k)}$ *are strictly positive.*

From relation [3.15], **P** is regular iff there exists an integer $k > 0$ such that all the elements of the *k*th power of **P** are strictly positive.

EXAMPLE 3.1.– i) If:

$$\mathbf{P} = \begin{bmatrix} .5 & .5 \\ 1 & 0 \end{bmatrix}, \qquad [3.21]$$

we have:

$$\mathbf{P}^2 = \begin{bmatrix} .75 & .25 \\ .5 & .5 \end{bmatrix} \qquad [3.22]$$

so that **P** is regular.

The transition graph associated with **P** is given in Figure 3.1.

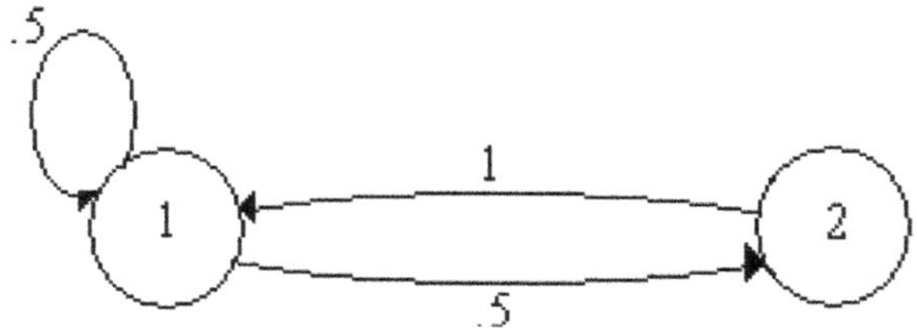

Figure 3.1. *Graph associated to matrix [3.21]*

ii) If:

$$\mathbf{P} = \begin{bmatrix} 1 & 0 \\ .75 & .25 \end{bmatrix}, \qquad [3.23]$$

P is not regular, because for any integer k,

$$p_{12}^{(k)} = 0. \qquad [3.24]$$

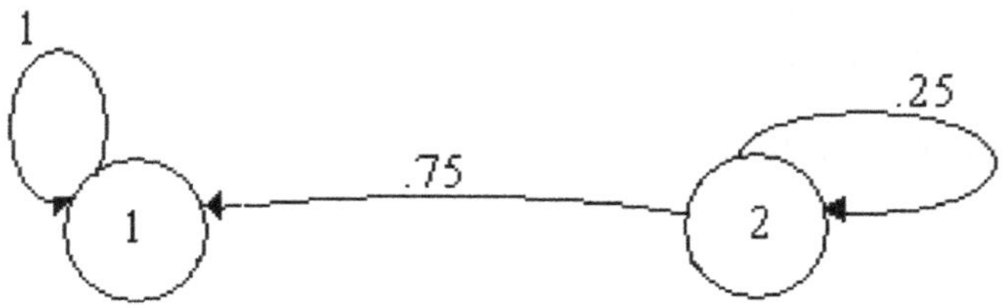

Figure 3.2. *Graph associated to matrix [3.23]*

The transition graph in this case is shown in Figure 3.2.

The same is true for the matrix:

$$\mathbf{P} = \begin{bmatrix} 0 & 1 \\ 1 & 0 \end{bmatrix}. \qquad [3.25]$$

iii) Any matrix **P** whose elements are all strictly positive is regular.

For example:

$$\begin{bmatrix} \frac{1}{3} & \frac{2}{3} \\ \frac{1}{4} & \frac{3}{4} \end{bmatrix} \qquad \begin{bmatrix} .7 & .2 & .1 \\ .6 & .2 & .2 \\ .4 & .1 & .5 \end{bmatrix}. \qquad [3.26]$$

3.2.2. *Markov chain state classification*

Let $i \in I$ and let $d(i)$ be the greatest common divisor of the set of integers n, such that:

$$p_{ii}^{(n)} > 0. \qquad [3.27]$$

DEFINITION 3.4.– *If* $d(i) > 1$, *the state* i *is said to be periodic with period* $d(i)$. *If* $d(i) = 1$, *then state* i *is aperiodic.*

Clearly, if $p_{ii} > 0$, then i is aperiodic. However, the converse is not necessarily true.

REMARK 3.1.– *If* **P** is regular, then all the states are aperiodic.

DEFINITION 3.5.– *A Markov chain whose states are all aperiodic is called an aperiodic Markov chain.*

From now on, we will have only Markov chains of this type.

DEFINITION 3.6.– *A state* i *is said to lead to state* j *(written* $i \triangleright j$*) iff there exists a positive integer n such that:*

$$p_{ij}^{n} > 0. \tag{3.28}$$

$i \mathrm{C} j$ *means that i does not lead to j.*

DEFINITION 3.7.– *States* i *and* j *are said to communicate iff* $i \triangleright j$ *and* $j \triangleright i$, *or if* $j = i$. *We write* $i \triangleleft\triangleright j$.

DEFINITION 3.8.– *A state* i *is said to be essential iff it communicates with every state it leads to; otherwise, it is called inessential.*

The relation $\triangleleft\triangleright$ defines an equivalence relation over the state space I resulting in a partition of I. The equivalence class containing state i is represented by $C(i)$.

DEFINITION 3.9.– *A Markov chain is said to be irreducible iff there exists only one equivalence class.*

Clearly, if **P** is regular, the Markov chain is both irreducible and aperiodic. Such a Markov chain is said to be *ergodic*.

It is easy to show that if the state i is essential (inessential), then all the elements of the class $C(i)$ are essential (inessential) (see [CHU 60]).

Thus, we can speak of essential and inessential classes.

DEFINITION 3.10.– *A subset* E *of the state space* I *is said to be closed iff:*

$$\sum_{j\in E} p_{ij} = 1, \textit{for all } i \in E. \qquad [3.29]$$

It can be shown that every essential class is minimally closed. See [CHU 60].

DEFINITION 3.11.– *For given states* i *and* j, *with* $J_0 = i$, *we can define the r.v.* τ_{ij} *called the first passage time to state j as follows:*

$$\tau_{ij} = \begin{cases} n & \text{if } J_\nu \neq j, \ \ 0 < \nu < n, \ \ J_n = j, \\ \infty & \text{if } J_\nu \neq j, \ \ \text{for all } \nu > 0. \end{cases} \qquad [3.30]$$

τ_{ij} is said to be the *hitting time* of the singleton $\{j\}$, starting from state i at time 0.

Supposing:

$$f_{ij}^{(n)} = P\left(\tau_{ij} = n \mid J_0 = i\right), \qquad n \in \mathbb{N}_0 \qquad [3.31]$$

and

$$f_{ij} = P\left(\tau_{ij} < \infty \mid J_0 = i\right), \qquad [3.32]$$

we can see that for $n > 0$:

$$f_{ij}^{(n)} = P(J_n = j, \ \ J_\nu \neq j, \ \ 0 < \nu < n \mid J_0 = i), \qquad [3.33]$$

$$= \sum_{S'_{n,i,j}} \prod_{k=0}^{n-1} p_{\alpha_k \alpha_{k+1}}, \qquad [3.34]$$

where the summation set $S'_{n,i,j}$ is defined as:

$$S'_{n,i,j} = \{(\alpha_0, \alpha_1, \ldots, \alpha_n) : \ \alpha_0 = i, \alpha_n = j, \alpha_k \in I, \alpha_k \neq j, \ \ k = 1, \ldots, n-1\}. \qquad [3.35]$$

We also have:

$$f_{ij} = \sum_{n=1}^{\infty} f_{ij}^{(n)}, \qquad [3.36]$$

$$1-f_{ij}=P\left(\tau_{ij}=\infty \mid J_0=i\right). \qquad [3.37]$$

The elements $f_{ij}^{(n)}$ can readily be computed by induction, using the following relations:

$$p_{ij}=f_{ij}^{(1)}, \qquad [3.38]$$

$$p_{ij}^{(n)}=\sum_{\nu=1}^{n-1} f_{ij}^{(\nu)} p_{jj}^{(n-\nu)}+f_{ij}^{(n)}, \quad n \geq 2. \qquad [3.39]$$

Let:

$$m_{ij}=E\left(\tau_{ij} \mid J_0=i\right), \qquad [3.40]$$

with the possibility of an infinite mean. The value of m_{ij} is given by:

$$m_{ij}=\sum_{n=1}^{\infty} n f_{ij}^{(n)}-\infty\left(1-f_{ij}\right).^{(*)} \qquad [3.41]$$

If $i=j$, then m_{ij} is called the *first passage time mean* or the *mean recurrence* time of state *i*.

For every j, we define the sequence of successive return times to state j $\left(r_n^{(j)},\ n \geq a\right)$ as follows:

$$r_0^{(j)}=0, \qquad [3.42]$$

$$r_n^{(j)}=\sup_k\left\{k \in \mathrm{N}_0,\ k>r_{n-1}^{(j)},\ J_\nu \neq j,\ r_{n-1}^{(j)}<\nu<k\right\},\ n>0. \qquad [3.43]$$

Using the Markov property and supposing $J_0=j$, the sequence of return times to state j is a renewal sequence with the r.v.

$$r_n^{(j)}-r_{n-1}^{(j)}, \quad n \geq 1 \qquad [3.44]$$

distributed according to τ_{jj}.

(*) Using the following conventions:
$\infty+a=\infty,\ a \in \mathbb{R},\ \infty \cdot a=\infty,\ (a>0)$, and in this particular case, $\infty \cdot 0=0$.

If $J_0 = i,\ \ i \neq j$, then $\left(r_n^{(j)},\ \ n \geq 0\right)$ is a delayed renewal sequence. In this case:

$$r_1^{(j)} = \tau_{ij}, \qquad [3.45]$$

and

$$r_n^{(j)} - r_{n-1}^{(j)} \sim \tau_{jj}, \qquad n > 1. \qquad [3.46]$$

This shows that a Markov chain contains many embedded renewal processes. These processes are used to define the next classification of states.

DEFINITION 3.12.– *A state i is said to be transient (recurrent) if the renewal process associated with its successive return times to i is transient (recurrent).*

A direct consequence of this definition is that:

$$i \text{ transient} \Leftrightarrow f_{ii} < 1, \qquad [3.47]$$

$$i \text{ recurrent} \Leftrightarrow f_{ii} = 1. \qquad [3.48]$$

A recurrent state i is said to be *null* (*positive*) if $m_{ii} = \infty \ \ \left(m_{ii} < \infty\right)$. It can be shown that if $m_{ii} < \infty$, then we can only have positive recurrent states.

This classification leads to the decomposition theorem (see [CHU 60]).

PROPOSITION 3.1 (DECOMPOSITION THEOREM).– The state space I of any Markov chain can be decomposed into r $(r \geq 1)$ subsets $C_1, \ldots, C_r$ forming a partition, such that each subset C_i is one and only one of the following types:

i) an essential recurrent positive closed set;

ii) an inessential transient non-closed set.

REMARK 3.2.–

1) If an inessential class reduces to a singleton $\{i\}$, there are two possibilities:

a) There exists a positive integer N such that:

$$0 < p_{ii}^{(N)} < 1. \qquad [3.49]$$

b) The N in a) does not exist. In this case, the state i is said to be a *non-return state*.

2) If the singleton $\{i\}$ forms an essential class, then:

$$p_{ii} = 1 \qquad [3.50]$$

and the state i is said to be an *absorbing state*.

The literature on Markov chains gives the following necessary and sufficient conditions for recurrence and transience.

PROPOSITION 3.2.–

i) State i is transient *iff*

$$\sum_{n=1}^{\infty} p_{ii}^{(n)} < \infty. \qquad [3.51]$$

In this case, for all $k \in I$:

$$\sum_{n=1}^{\infty} p_{ki}^{(n)} < \infty, \qquad [3.52]$$

and in particular:

$$\lim_{n \to \infty} p_{ki}^{(n)} = 0, \quad \forall k \in I. \qquad [3.53]$$

ii) State i is recurrent *iff*

$$\sum_{n=1}^{\infty} p_{ii}^{(n)} = \infty. \qquad [3.54]$$

In this case:

$$k \lhd\rhd i \Rightarrow \sum_{n=1}^{\infty} p_{ki}^{(n)} = \infty, \qquad [3.55]$$

and

$$k \, C \, i \Rightarrow \sum_{n=1}^{\infty} p_{ki}^{(n)} = 0. \qquad [3.56]$$

3.2.3. *Computation of absorption probabilities*

PROPOSITION 3.3.–

i) If i is recurrent and if $j \in C(i)$, then $f_{ij} = 1$.

ii) If i is recurrent and if $j \notin C(i)$, then $f_{ij} = 0$.

PROPOSITION 3.4.– Let T be the set of all transient states of I, and let C be a recurrent class.

For all $j,\ k \in C$,

$$f_{ij} = f_{ik}.$$

Labeling this common value as f_{iC}, the probabilities $\left(f_{i,C},\ i \in T\right)$ satisfy the linear system:

$$f_{i,C} = \sum_{k \in T} p_{ik} f_{k,C} + \sum_{k \in C} p_{ik}, \qquad i \in T. \tag{3.57}$$

REMARK 3.3.– Parzen [PAR 62] proved that under the assumption of proposition 3.4, the linear system [3.57] has a unique solution. This shows, in particular, that if there is only one irreducible class C, then for all $i \in T$:

$$f_{i,C} = 1. \tag{3.58}$$

DEFINITION 3.13.– *The probability* $f_{i,C}$ *introduced in proposition 3.4 is called absorption probability in class C, starting from state i.*

If class C is recurrent:

$$f_{i,C} = \begin{cases} 1 & \text{if } i \in C, \\ 0 & \text{if } i \text{ is recurrent, } i \notin C. \end{cases} \tag{3.59}$$

3.2.4. *Asymptotic behavior*

Consider an irreducible aperiodic Markov chain which is positive recurrent.

Suppose that the following limit exists:

$$\lim_{n\to\infty} p_j(n) = \pi_j, \quad j \in I \tag{3.60}$$

starting with $J_0 = i$.

The relation:

$$p_j(n+1) = \sum_{k\in I} p_k(n) p_{kj} \tag{3.61}$$

becomes:

$$p_{ij}^{(n+1)} = \sum_{k\in I} p_{ik}^{(n)} p_{kj}, \tag{3.62}$$

because

$$p_j(n) = p_{ij}^{(n)}. \tag{3.63}$$

Since the state space I is finite, we obtain from [3.60] and [3.62]:

$$\pi_j = \sum_{k\in I} \pi_k p_{kj}, \tag{3.64}$$

and from [3.63]:

$$\sum_{i\in I} \pi_i = 1. \tag{3.65}$$

The result:

$$\lim_{n\to\infty} p_{ij}^{(n)} = \pi_j \tag{3.66}$$

is called an *ergodic result*, since the value of the limit in [3.66] is independent of the initial state i.

From result [3.66] and [3.17], we see that for any initial distribution **p**:

$$\lim_{n\to\infty} p_i(n) = \lim_{n\to\infty} \sum_j p_j p_{ji}^{(n)} \quad [3.67]$$

$$= \sum_j p_j \pi_i , \quad [3.68]$$

so that:

$$\lim_{n\to\infty} p_i(n) = \pi_i . \quad [3.69]$$

This shows that the asymptotic behavior of a Markov chain is given by the existence (or non-existence) of the limit of the matrix $\mathbf{P}^n$.

A standard result concerning the asymptotic behavior of $\mathbf{P}^n$ is given in the next proposition. The proof can be found in [CHU 60], [PAR 62] or [FEL 57].

PROPOSITION 3.5.– For any aperiodic Markov chain of transition matrix **P** and having a finite number of states, we have:

a) if state j is recurrent (necessarily positive), then:

i) $i \in C(j) \Rightarrow \lim_{n\to\infty} p_{ij}^{(n)} = \dfrac{1}{m_{jj}}$, [3.70]

ii) i recurrent and $\notin C(j) \Rightarrow \lim_{n\to\infty} p_{ij}^{(n)} = 0$, [3.71]

iii) i transient $\lim_{n\to\infty} p_{ij}^{(n)} = \dfrac{f_{i,C(j)}}{m_{jj}}$. [3.72]

b) If j is transient, then for all $i \in I$:

$$\lim_{n\to\infty} p_{ij}^{(n)} = 0. \quad [3.73]$$

REMARK 3.4.–

Result ii) of part a) is trivial since in this case:

$p_{ij}^{(n)} = 0$ for all positive n.

From proposition 3.5, the following corollaries can be deduced.

COROLLARY 3.1 (IRREDUCIBLE CASE).– If the Markov chain of transition matrix **P** is irreducible, then for all $i, j \in I$:

$$\lim_{n\to\infty} p_{ij}^{(n)} = \pi_j, \quad [3.74]$$

with

$$\pi_j = \frac{1}{m_{jj}}. \quad [3.75]$$

It follows that for all j:

$$\pi_j > 0. \quad [3.76]$$

If we use remark 3.3 in the particular case where we have only one recurrent class and where the states are transient (the so-called *uni-reducible* case), then we have the following corollary.

COROLLARY 3.2 (UNI-REDUCIBLE CASE).– If the Markov chain of transition matrix **P** has one essential class C (necessarily recurrent positive) and T as transient set, then we have:

i) for all $i,\ j \in C$: $\lim_{n\to\infty} p_{ij}^{(n)} = \pi_j$, [3.77]

with $\{\pi_j, j \in C\}$ being the unique solution of the system:

$$\pi_j = \sum_{i\in C} \pi_i p_{ij}, \quad [3.78]$$

$$\sum_{j\in C} \pi_j = 1. \quad [3.79]$$

ii) for all $j \in T$:

$$\lim_{n\to\infty} p_{ij}^{(n)} = 0 \text{ for all } i \in I. \quad [3.80]$$

iii) for all $j \in C$:

$$\lim_{n\to\infty} p_{ij}^{(n)} = \pi_j \text{ for all } i \in T. \quad [3.81]$$

REMARK 3.5.– Relations [3.79] and [3.77] are true because the set C of recurrent states can be seen as a Markov subchain of the initial chain.

If the ℓ transient states belong to the set $\{1,\ldots,\ell\}$, using a permutation of the set I, if necessary, then the matrix **P** takes the following form:

$$\begin{array}{c} \\ 1 \\ \vdots \\ \ell \\ \ell+1 \\ \vdots \\ m \end{array} \begin{array}{c} \begin{array}{cccccc} 1 & \cdots & \ell & \ell+1 & \cdots & m \end{array} \\ \left[\begin{array}{cc} \mathbf{P}_{11} & \mathbf{P}_{12} \\ & \\ \mathbf{O} & \mathbf{P}_{22} \end{array}\right] \end{array} = \mathbf{P}. \quad [3.82]$$

This proves that the submatrix $\mathbf{P}_{22}$ is itself a Markov transition matrix.

Let us now consider a Markov chain of matrix **P**. The general case is given by a partition of I:

$$I = T \cup C_1 \cup \cdots \cup C_r, \quad [3.83]$$

where T is the set of transient states and $C_1,\ldots,C_r$ are the r positive recurrent classes.

By reorganizing the order of the elements of I, we can always suppose that:

$$T = \{1,\ldots,\ell\}, \quad [3.84]$$

$$C_1 = \{\ell+1,\ldots,\ell+\nu_1\}, \quad [3.85]$$

$$C_2 = \{\ell+\nu_1+1,\ldots,\ell+\nu_1+\nu_2\}, \quad [3.86]$$

$$\vdots$$

$$C_r = \left\{\ell+\sum_{j=1}^{r-1}\nu_j+1,\ldots,\ m\right\}, \quad [3.87]$$

where ν_j is the number of elements in C_j, $(j=1,\ldots,r)$ and

$$\ell+\sum_{j=1}^{r}\nu_j=m. \qquad [3.88]$$

This results from the following block partition of matrix **P**:

$$\mathbf{P}=\begin{bmatrix} \mathbf{P}_{\ell\times\ell} & \mathbf{P}_{\ell\times\nu_1} & \mathbf{P}_{\ell\times\nu_2} & \cdots & \mathbf{P}_{\ell\times\nu_r} \\ \mathbf{0} & \mathbf{P}_{\nu_1\times\nu_1} & \mathbf{0} & \cdots & \mathbf{0} \\ \mathbf{0} & \mathbf{0} & \mathbf{P}_{\nu_2\times\nu_2} & \cdots & \mathbf{0} \\ \vdots & \vdots & \vdots & \ddots & \vdots \\ \mathbf{0} & \mathbf{0} & \mathbf{0} & \cdots & \mathbf{P}_{\nu_r\times\nu_r} \end{bmatrix} \qquad [3.89]$$

where, for $j=1,\ldots, r$:

$\mathbf{P}_{\ell\times\ell}$ is the transition submatrix for T.

$\mathbf{P}_{\ell\times\nu_j}$ is the transition submatrix from T to C_j.

$\mathbf{P}_{\nu_j\times\nu_j}$ is the transition submatrix for the class C_j.

From proposition 3.1, we have the following corollary:

COROLLARY 3.3.– For a general Markov chain of matrix **P**, given by [3.89], we have:

i) For all $i\in I$ and all $j\in T$: $\lim_{n\to\infty} p_{ij}^{(n)}=0.$ [3.90]

ii) For all $j\in C_\nu$ $\quad (\nu=1,\ldots, r)$:

$$\lim_{n\to\infty} p_{ij}^{(n)}=\begin{cases}\pi_j & \text{if } i\in C_\nu, \\ 0 & \text{if } i\in C_{\nu'} \quad \nu'\neq\nu, \\ f_{i,C_\nu}\pi_j^\nu & \text{if } i\in T.\end{cases} \qquad [3.91]$$

Moreover, for all $\nu=1,\ldots, r$:

$$\sum_{j\in C_\nu}\pi_j^\nu=1. \qquad [3.92]$$

This last result allows us to calculate the limit values quite simply.

For $\left(\pi_j^\nu,\ j\in C_\nu\right),\ \nu=1,\ldots,\ r$, it suffices to solve the linear systems for each fixed ν:

$$\begin{cases}\pi_j^\nu=\sum_{k\in C_\nu}\pi_k^\nu p_{kj}, & j\in C_\nu,\\ \sum_{i\in C_\nu}\pi_i^\nu=1.\end{cases}\qquad [3.93]$$

Indeed, since each C_ν is itself a space set of an irreducible Markov chain of matrix $\mathbf{P}_{\nu\times\nu}$, the above relations are none other than [3.78] and [3.79].

For the absorption probabilities $\left(f_{i,C_\nu},\ i\in T\right),\ \nu=1,\ldots,\ r$, it suffices to solve the following linear system for each fixed ν. Using relation [3.54], we have:

$$f_{i,C_\nu}=\sum_{k\in T}p_{ik}f_{i,C_\nu}+\sum_{k\in C_\nu}p_{ik},\qquad i\in T.\qquad [3.94]$$

An algorithm given in [DED 84] is very useful for the classification of the states of a Markov chain. It is fully developed in [JAN 06].

3.2.5. *Example: a management problem in an insurance company*

A car insurance company classifies its customers into three groups:

– G_0 : those having no accidents during the year;

– G_1 : those having one accident during the year;

– G_2 : those having more than one accident during the year.

The statistics department of the company observes that the annual transition between the three groups can be represented by a Markov chain with state space $\{G_0,G_1,G_2\}$ and transition matrix $\mathbf{P}$:

$$\mathbf{P}=\begin{bmatrix}.85 & .10 & .05\\ 0 & .80 & .20\\ 0 & 0 & 1\end{bmatrix}.\qquad [3.95]$$

We suppose that the company produces 50,000 new contracts per year and wants to know the distribution of these contracts for the next 4 years.

After 1 year, we have, on average:

– in group G_0 : $50,000 \times .85 = 42,500$;

– in group G_1 : $50,000 \times .10 = 5,000$;

– in group G_2 : $50,000 \times .05 = 2,500$.

These results are simply the elements of the first row of **P**, multiplied by 50,000. After 2 years, multiplying the elements of the first row of $\mathbf{P}^{(2)}$ by 50,000, we obtain:

– in group G_0 : $36,125$;

– in group G_1 : $8,250$;

– in group G_2 : $5,625$.

A similar computation gives the following results.

	After 3 years	After 4 years
G_0	30,706	26,100
G_1	10,213	11,241
G_3	9,081	12,659

To find the type of the Markov chain with transition matrix (9.109), the simple graph of possible transitions given in Figure 3.3 shows that the class $\{1,\ 2\}$ is transient and class $\{3\}$ is absorbing. Thus, using corollary 3.2, we obtain the limit matrix:

$$\mathbf{A} = \begin{bmatrix} 0 & 0 & 1 \\ 0 & 0 & 1 \\ 0 & 0 & 1 \end{bmatrix}. \qquad [3.96]$$

The limit matrix can be interpreted as showing that regardless of the initial composition of the group, the customers will finish by having at least two accidents.

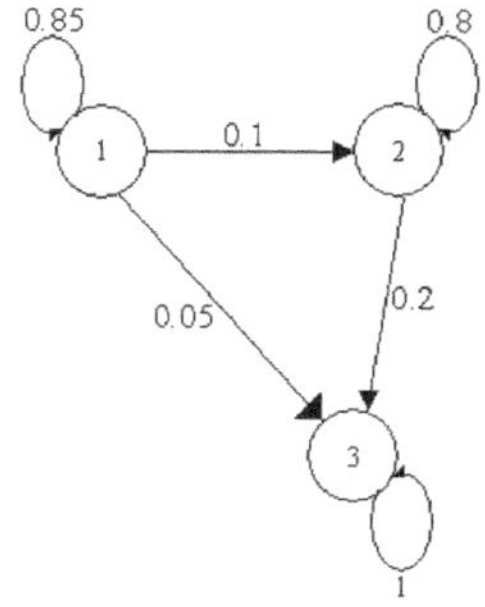

Figure 3.3. *Graph of the states*

REMARK 3.6.– If we want to know the situation after one or two changes, we can use relation [1.19] with $n = 1,2,3$ and with $\mathbf{p}$ given by:

$$\mathbf{p} = (.26,.60,.14). \qquad [3.97]$$

We obtain the following results:

$$\begin{array}{lll} p_1^{(1)} = .257 & p_2^{(1)} = .597 & p_3^{(1)} = .146 \\ p_1^{(2)} = .255 & p_2^{(2)} = .594 & p_3^{(2)} = .151 \\ p_1^{(3)} = .254 & p_2^{(3)} = .590 & p_3^{(3)} = .156. \end{array}$$

These results show that the convergence of $\mathbf{p}^{(n)}$ to π is relatively fast.

3.3. Non-homogeneous Markov chains

3.3.1. *Definitions*

DEFINITION 3.14.– *The random sequence* $(J_n,\ n \in \mathrm{N})$ *is a non-homogeneous Markov chain iff for all* $j_0,\ j_1,\ldots,j_n \in I, \forall n \in \mathbb{N}_0$:

$$P(J_n = j_n \mid J_0 = j_0,\ J_1 = j_1,\ldots,\ J_{n-1} = j_{n-1}) = P(J_n = j_n \mid J_{n-1} = j_{n-1}) \qquad [3.98]$$

(provided that this probability has meaning) where the probability of the second member depends on n, i.e.:

$$P(J_n = j \mid J_{n-1} = i) = p_{ij}(n). \qquad [3.99]$$

With matrix notations, we write:

$$\mathbf{P(n)} = (p_{ij}(n)).$$

It follows that a *non-homogeneous Markov chain* is defined by vector [3.99] with conditions [3.100]:

$$\mathbf{p} = \left(p_1, \ldots, p_m\right), \tag{3.100}$$

such that:

$$\begin{aligned} &p_i \geq 0, \qquad i \in I, \\ &\sum_{i \in I} p_i = 1 \end{aligned} \tag{3.101}$$

and a sequence of *mxm* Markov matrices $\Pi = (\mathbf{P(n)}, n \in \mathbb{N}_0)$ or shorter by the couple $(\mathbf{p}, \Pi)$.

We now introduce the transition probabilities of order $p_{ij}^{(n)}$, defined as:

$$p_{ij}^{(n)}(\nu) = P\left(J_{\nu+n} = j \mid J_\nu = i\right), n \in \mathbb{N}_0, \nu \in \mathbb{N}. \tag{3.102}$$

also noted

$$p_{ij}^{(n)}(\nu) = p_{ij}(\nu, \nu + n)$$

or [3.103]

$$p_{ij}^{(n)}(\nu) = p_{ij}(s,t) \text{ with } \nu = s, t = \nu + n.$$

With matrix notations, we write:

$$\mathbf{P}(\nu, \nu + n) = (p_{ij}^{(n)}(\nu))$$

or [3.104]

$$\mathbf{P}(\nu, \nu + n) = \mathbf{P}(s,t) \text{ with } \nu = s, t = \nu + n.$$

From the Markov property [3.98], it is clear that conditioning with respect to $J_{\nu+1}$, we obtain:

$$p_{ij}(0,2) = \sum_k p_{ik}(0,1 p_{kj}(1,2). \tag{3.105}$$

or

$$\mathbf{P}(0,2)=\mathbf{P}(0,1)\mathbf{P}(1,2). \quad [3.106]$$

These last two relations can be extended first to:

$$\begin{aligned} p_{ij}(0,n) &= \sum_k p_{ik}(0,n-1)p_{kj}(n-1,n), \\ \mathbf{P}(0,n) &= \mathbf{P}(0,n-1)\mathbf{P}(n-1,n). \end{aligned} \quad [3.107]$$

By recurrence, finally, we have:

$$\mathbf{P}(0,n)=\mathbf{P}(0,1)\mathbf{P}(1,2)\ldots\mathbf{P}(n-1,n). \quad [3.108]$$

This last relation extends relation [3.15] seen in the homogeneous case.

As in relation [3.16], the marginal distributions related to J_n and $n \geq 0$ are given by:

$$p_i(n)=P(J_n=i). \quad [3.109]$$

These probabilities may be computed as follows:

$$p_i(n)=\sum_j p_j p_{ji}(0,n), \quad i \in I. \quad [3.110]$$

If we write:

$$p_{ji}^{(0)}=\delta_{ji} \text{ or } \mathbf{P}^{(0)}=\mathbf{I}, \quad [3.111]$$

then relation [3.15] is true for all $n \geq 0$.

If:

$$\mathbf{p}(n)=\left(p_1(n),\ldots,p_m(n)\right), \quad [3.112]$$

then relation [3.110] can be expressed, using matrix notation, as:

$$\mathbf{p}(n)=\mathbf{p}\mathbf{P}(0,n). \quad [3.113]$$

It is also possible to start at time s with an initial distribution $\mathbf{p}$ for state J_s. Then, if:

$$\begin{aligned} &p_i(s,t) = P_{\mathbf{p}}(J_t = i), i \in I, \\ &\mathbf{p}(s,t) = (p_1(s,t),...,p_m(s,t)) \end{aligned}, \qquad [3.114]$$

We have as an extension of result [3.113]

$$\mathbf{p}(s,t) = \mathbf{p}\mathbf{P}(s,s+1)...\mathbf{P}(s+t-1,t). \qquad [3.115]$$

REMARK 3.7.– Partially non-homogeneous Markov chains

In some applications, it may be possible that after a certain time T, the successive changes of states follow a homogeneous Markov chain of matrix $\mathbf{P}$.

If so, relation [3.108] becomes:

$$\mathbf{P}(\mathbf{0},n) = \begin{cases} \mathbf{P}(\mathbf{0},1)\mathbf{P}(1,2)...\mathbf{P}(n-1,n), n \le N \\ \mathbf{P}(\mathbf{0},1)\mathbf{P}(1,2)...\mathbf{P}(n-1,n)\mathbf{P}^m, n = N+m. \end{cases} \qquad [3.116]$$

In this case, the Markov chain will be called *a partially non-homogeneous Markov chain or delayed Markov chain of order N.*

3.3.2. *Asymptotical results*

As we have already mentioned for non-homogeneous renewal processes, there are not many useful theoretical results which we can be used for application particularly for the asymptotical behavior. This is why we will develop numerical methods in the next section. Nevertheless, some results exist and they can be found, for example, in [ISA 76] in which the concepts of *strong* and *weak ergodicity* are introduced.

For example, *strong ergodicity* means that there exists a vector $\boldsymbol{\pi} = (\pi_1,...,\pi_m)$,

with: $\pi_i \ge 0, i = 1,...,m$ and $\sum_{i=1}^{m} p_i = 1$

such that: $\lim_{t\to\infty} \sup_{\mathbf{p}} \|\mathbf{p}(s,t) - \boldsymbol{\pi}\| = 0$.

Of course in the particular case of a partially non-homogeneous Markov chain, the s asymptotical behavior is given by one of the homogeneous Markov chains of matrix $\mathbf{P}$.

3.4. Markov reward processes

3.4.1. *Classification and notation*

3.4.1.1. *Classification of reward processes*

The association of a sum of money with a state of the system and with a state transition assumes great relevance. In general, this can be done by attaching a reward structure to the process.

This structure can be seen as a random variable associated with the state occupancies and transitions [HOW 71].

The rewards can be of different kinds, but in this book we will, because of our kind of applications, consider only amounts of money. These amounts can be positive, if for the system they can be seen as a *benefit* and negative if they can be considered as a *cost*.

The reward processes can be seen as classes of stochastic processes that we can classify into different cases. The following tables report the classification of the reward processes.

Process classification

<table>
<tr><td colspan="3">Homogeneous</td></tr>
<tr><td colspan="3">Non-homogeneous</td></tr>
<tr><td colspan="3"></td></tr>
<tr><td colspan="3">Continuous time</td></tr>
<tr><td colspan="3">Discrete time</td></tr>
<tr><td colspan="3"></td></tr>
<tr><td colspan="3">Non-discounted</td></tr>
<tr><td rowspan="3">Discounted</td><td colspan="2">Fixed interest rate</td></tr>
<tr><td rowspan="2">Variable interest rate</td><td>Homogeneous interest law</td></tr>
<tr><td>Non-homogeneous interest law</td></tr>
</table>

Reward classification

<table>
<tr><td colspan="3">Time fixed rewards</td></tr>
<tr><td rowspan="2">Time variable rewards</td><td colspan="2">Homogeneous rewards</td></tr>
<tr><td colspan="2">Non-homogeneous rewards</td></tr>
<tr><td colspan="3"></td></tr>
<tr><td colspan="3">Transition (impulse) rewards</td></tr>
<tr><td rowspan="5">Permanence (rate) rewards</td><td rowspan="2">Discrete time</td><td>Immediate</td></tr>
<tr><td>Due</td></tr>
<tr><td colspan="2"></td></tr>
<tr><td colspan="2">Independent on next transition</td></tr>
<tr><td colspan="2">Dependent on next transition</td></tr>
</table>

We will not present permanence rewards that depend on the next transition because in financial and insurance environments they do not make sense.

In general, the distinction between homogeneous and non-homogeneous cases is done for stochastic processes. Also, an interest rate law can be defined as homogeneous if the discount factor is a function only of the length of the financial operation, non-homogeneous if the discount factor takes also into account the initial time of operation, and not only the duration.

In the same way, rewards can be fixed in time, can depend only on the duration or can be non-homogeneous in time.

We will use the following notation:

$\psi_i, \psi_i(t), \psi_i(s,t)$: represent rewards given for permanence in state *i;* the first is time fixed, the second changes because of time and the third represents a time non-homogeneous permanence reward;

$\gamma_{ij}, \gamma_{ij}(t), \gamma_{ij}(s,t)$: represent the three different kinds of rewards given for the transition from state *i* to state *j* (impulse reward).

In the discrete time case, the *immediate case* means that the reward is paid at the end of each period; in the *due case,* the reward is paid at the beginning of the period. The impulse rewards γ represent lump sums that are paid at the transition instant.

3.4.1.2. *Financial parameters*

To study the process with discounting, let us recall some basic results for computing present values of amounts of money, annuities and also the related notation.

For more details, refer to [JAN 09b].

i) Fixed time interest rate

$v = (1+r)^{-1} = e^{-\delta}$: represents the one-period discount factor, where r is the interest rate and δ is the corresponding intensity;

$a_{\overline{t|}r} = \dfrac{1-v^t}{r}$; $\ddot{a}_{\overline{t|}r} = \dfrac{1-v^t}{d}, d = \dfrac{r}{1+r}$: represent the present value of, respectively, a unitary annuity-immediate and an annuity-due;

$a_{\overline{\infty|}r} = \dfrac{1}{r}, \ddot{a}_{\overline{\infty|}r} = \dfrac{1}{d}$: represents the present value of infinite time unitary annuities, also called *perpetuities.*

ii) Time variable interest rate

Now, we suppose that the interest rates are variable and depend on the time period:

$v(k) = \prod_{h=1}^{k} (1+r_h)^{-1}$ represents the *k-period discount factor* at time 0 in discrete time. It gives, at time 0, the value of one monetary unit to be paid at the end of period k;

$\sum_{k=1}^{t} v(k)\psi_i(k), \sum_{k=0}^{t-1} v(k)\psi_i(k)$ represent the present value, respectively, of an annuity-immediate and an annuity-due with variable rewards in the time and variable interest rate on a time horizon t.

The infinite rate cases are given by the limit to ∞ of the two relations. They converge depending on the values of ψ.

$v(s,t) = \prod_{h=s+1}^{t} (1+r_h)^{-1}$ represents, in the non-homogeneous case, the $t-s$ *period discount factor at times*, with homogeneous interest rates giving, at time s, the discounted value of one monetary unit to be paid at the end of period k in discrete time;

$\sum_{k=s+1}^{t} v(s,k)\psi_i(s,k), \sum_{k=s}^{t-1} v(s,k)\psi_i(s,k)$ represent the present value at time s, respectively, of an annuity-immediate and an annuity-due paid on the time interval $s, s+1, \ldots, t$ with non-homogeneous rewards and variable interest rate;

$\dot{v}(s,t) = \prod_{h=s+1}^{t} (1+ {}_s r_h)^{-1}$ represents the $t-s$ period discount factors at time s, with non-homogeneous interest rates;

$\sum_{k=s+1}^{t} \dot{v}(s,k)\psi_i(s,k), \sum_{k=s}^{t-1} \dot{v}(s,k)\psi_i(s,k)$ represent the present value at time s, respectively, of an annuity-immediate and an annuity-due paid on the time interval $s, s+1, \ldots, t$ with non-homogeneous rewards and non-homogeneous interest rate.

In the following sections, and in Chapter 4, we will show that annuities are strongly related to the Markov and semi-Markov reward processes.

A reward structure can be considered as a very general structure that, given a financial and economic meaning, can be very useful in stochastic modeling.

For example, this behavior is particularly efficient to construct models useful to follow the dynamic evolution of insurance problems. In this case, the permanence in a state usually involves the payment of a premium or the receipt of a claim. Furthermore, often the transition from one state to another induces some other costs or benefits.

3.5. Discrete time Markov reward processes (DTMRWPs)

3.5.1. *Undiscounted case*

Suppose that we have a system which evolves randomly within the Markov property having $I = \{1, \ldots, m\}$ as state space. This means, as we have already discussed, that it moves in such a way that *the future is a function only of the present*. Moreover, a reward structure is attached to the evolution of the system.

3.5.1.1. *First model*

In our first model, we suppose that:

1) rewards are fixed in time;

2) rewards are given only for permanence in the state;

3) rewards are paid at the end of each period.

Under these assumptions, if $V_i^{(n)}$ represents the mean total rewards (MTRs) paid or received after n transitions, given that at time 0 the system was in the state i, then in the homogeneous case, the following results are obtained:

$$V_i^{(0)} = 0, \quad [3.117]$$

$$V_i^{(1)} = \psi_i = V_i^{(0)} + \psi_i , \quad [3.118]$$

$$V_i^{(2)} = 0 + \psi_i + \sum_{k=1}^{m} p_{ik}^{(1)} \psi_k = V_i^{(1)} + \sum_{k=1}^{m} p_{ik}^{(1)} \psi_k , \quad [3.119]$$

and in general:

$$V_i^{(n)} = V_i^{(n-1)} + \sum_{k=1}^{m} p_{ik}^{(n-1)} \psi_k . \quad [3.120]$$

For the non-homogeneous case, the preceding results become:

$$V_i^{(n)}(s) = V_i^{(n-1)}(s) + \sum_{k=1}^{m} p_{ik}^{(n-1)}(s) \psi_k . \quad [3.121]$$

Relations [3.120] and [3.121] give the MTRs that will be expected at time n if the system is in state i at time s or at time $s = 0$ for the homogeneous case. In both cases, these results are obtained recursively adding the MTR expected at time $(s + n - 1)$ to the mean reward that will be paid at time $(s + n)$.

3.5.1.2. *Second model*

For our second model, we assume that:

1) rewards are variable in time;

2) rewards are given only for permanence in the state;

3) rewards are paid at the end of the period.

In this case, for the homogeneous case, we obtain:

$$V_i^{(1)} = \psi_i(1) , \quad [3.122]$$

$$V_i^{(2)} = V_i^{(1)} + \sum_{k=1}^{m} p_{ik}^{(1)} \psi_k(2) , \quad [3.123]$$

$$V_i^{(n)} = V_i^{(n-1)} + \sum_{k=1}^{m} p_{ik}^{(n-1)} \psi_k(n) . \quad [3.124]$$

For the non-homogeneous case, this general result becomes:

$$V_i^{(n)}(s) = V_i^{(n-1)}(s) + \sum_{k=1}^{m} p_{ik}^{(n-1)}(s)\psi_k(s+n). \qquad [3.125]$$

Recursive relations [3.124] and [3.125] for MTR have the same meaning as relations [3.120] and [3.121].

3.5.1.3. *Third model*

Now, we suppose that:

1) rewards are variable in time;

2) rewards are given for permanence in the state and at a given transition;

3) rewards are paid at the end of the period.

For the homogeneous case, the following formulas are obtained:

$$V_i^{(1)} = \psi_i(1) + \sum_{k=1}^{m} p_{ik}\gamma_{ik}(1), \qquad [3.126]$$

$$V_i^{(n)} = V_i^{(n-1)} + \sum_{k=1}^{m} p_{ik}^{(n-1)}\left(\psi_k(n) + \sum_{j=1}^{m} p_{kj}\gamma_{kj}(n)\right), \qquad [3.127]$$

and for the non-homogeneous case:

$$V_i^{(n)}(s) = V_i^{(n-1)}(s) + \sum_{k=1}^{m} p_{ik}^{(n-1)}(s)\left(\psi_k(s+n) + \sum_{j=1}^{m} p_{kj}(s+n)\gamma_{kj}(s+n)\right). \qquad [3.128]$$

Recursive relations [3.127] and [3.128] outline the presence in the MTR of the lump sums that are given or taken at the instant of the change of state.

3.5.1.4. *Fourth model*

The last immediate case considers non-homogeneous rewards.

We suppose that:

1) rewards vary because of the starting time and the time spent after (non-homogeneous rewards);

2) rewards are also given at the transitions;

3) rewards are paid at the end of the period.

In this case, the general formula is:

$$V_i^{(n)}(s) = V_i^{(n-1)}(s) + \sum_{k=1}^{m} p_{ik}^{(n-1)}(s)\left(\psi_k(s,s+n) + \sum_{j=1}^{m} p_{kj}(s+n)\gamma_{kj}(s,s+n)\right). \qquad [3.129]$$

In the undiscounted case, the difference between the due and immediate cases is not relevant.

3.5.2. *Discounted case*

3.5.2.1. *Immediate cases*

In our system, any time that a reward is paid it will be discounted at time 0 in the homogeneous case and at time s in the non-homogeneous case.

3.5.2.1.1. *First model*

In the first formulation of this case, we suppose that:

1) rewards are fixed in time;

2) rewards are given only for permanence in the state;

3) rewards are paid at the end of the period;

4) the interest rate per period r is fixed.

In this case, $V_i^{(n)}$ represents the *reward mean present value* (RWMPV) of all the rewards paid or received after n transitions, given that at time 0 the system was in the state i and $V_i^{(n)}(s)$ represents the RWMPV of all the rewards paid or received after n transitions, given that at time s the system was in the state i.

Under these assumptions, we obtain for the homogeneous case:

$$V_i^{(1)} = (1+r)^{-1}\psi_i = (1+r)^{-1}\psi_i, \qquad [3.130]$$

$$V_i^{(2)} = (1+r)^{-1}\psi_i + v^2\sum_{k=1}^{m} p_{ik}^{(1)}\psi_k = V_i^{(1)} + v^2\sum_{k=1}^{m} p_{ik}^{(1)}\psi_k, \qquad [3.131]$$

$$V_i^{(n)} = V_i^{(n-1)} + v^n\sum_{k=1}^{m} p_{ik}^{(n-1)}\psi_k, \qquad [3.132]$$

and for the non-homogeneous case:

$$V_i^{(n)}(s) = V_i^{(n-1)}(s) + v^n \sum_{k=1}^{m} p_{ik}^{(n-1)}(s)\psi_k . \qquad [3.133]$$

Formulas [3.132] and [3.133] give the RWMPV in the simplest homogeneous and non-homogeneous cases, i.e. with a fixed interest rate and fixed rewards. We note that the introduction of interest rates preserves the recursivity of the formulas.

3.5.2.1.2. Second model

Now, we introduce the variable interest rates and will use the formulas given in the first paragraph of this section for discounting.

In this first formulation with variable interest rate, we have that:

1) rewards are fixed in time;

2) rewards are given only for permanence in the state;

3) rewards are paid at the end of the period;

4) the interest rate r is deterministic but variable with time.

In this case, the values of $V_i^{(n)}$ and $V_i^{(n)}(s)$ (RWMPV) will be obtained taking into account the deterministic interest rate structure. Under these hypotheses, we get the following formulas for the homogeneous case:

$$V_i^{(2)} = v(1)\psi_i + v(2)\sum_{k=1}^{m} p_{ik}^{(1)}\psi_k = V_i^{(1)} + v(2)\sum_{k=1}^{m} p_{ik}^{(1)}\psi_k , \qquad [3.134]$$

$$V_i^{(n)} = V_i^{(n-1)} + v(n)\sum_{k=1}^{m} p_{ik}^{(n-1)}\psi_k , \qquad [3.135]$$

and for the non-homogeneous case:

$$V_i^{(n)}(s) = V_i^{(n-1)}(s) + v(s, s+n)\sum_{k=1}^{m} p_{ik}^{(n-1)}(s)\psi_k . \qquad [3.136]$$

Relations [3.135] and [3.136] give the RWMPV in homogeneous and non-homogeneous cases with fixed rewards and variable interest rates. Here too the recursivity is preserved.

3.5.2.1.3. Third model

Now, we introduce the variability of rewards with the following hypotheses:

1) rewards are variable in time;

2) rewards are given only for permanence in the state;

3) rewards are paid at the end of the period;

4) the interest rate r is fixed.

In the homogeneous case, the following formulas, still recursive, hold:

$$V_i^{(2)} = V_i^{(1)} + v^2 \sum_{k=1}^{m} p_{ik}^{(1)} \psi_k(2), \quad [3.137]$$

$$V_i^{(n)} = V_i^{(n-1)} + v^n \sum_{k=1}^{m} p_{ik}^{(n-1)} \psi_k(n), \quad [3.138]$$

and in the non-homogeneous case:

$$V_i^{(n)}(s) = V_i^{(n-1)}(s) + v^n \sum_{k=1}^{m} p_{ik}^{(n-1)}(s) \psi_k(s+n). \quad [3.139]$$

3.5.2.1.4. Fourth model

Now, we consider the case of variable interest rates with variable rewards under the following assumptions:

1) rewards are variable in time;

2) rewards are given only for permanence in the state;

3) rewards are paid at the end of the period;

4) the interest rates are time-dependent.

For the homogeneous case, related recursive formulas for RWMPV are:

$$V_i^{(2)} = V_i^{(1)} + v(2) \sum_{k=1}^{m} p_{ik}^{(1)} \psi_k(2), \quad [3.140]$$

$$V_i^{(n)} = V_i^{(n-1)} + v(n) \sum_{k=1}^{m} p_{ik}^{(n-1)} \psi_k(n), \quad [3.141]$$

and for the non-homogeneous case:

$$V_i^{(n)}(s) = V_i^{(n-1)}(s) + v(s, s+n) \sum_{k=1}^{m} p_{ik}^{(n-1)}(s) \psi_k(s+n). \qquad [3.142]$$

3.5.2.1.5. Fifth model

Now, we will introduce the γ rewards under the following assumptions:

1) rewards are variable in time;

2) rewards are given for permanence in the state and at a given transition;

3) rewards are paid at the end of the period;

4) the interest rate r is fixed.

Under these hypotheses, for homogeneous and non-homogeneous cases, respectively, we obtain the following recursive relations:

$$V_i^{(n)} = V_i^{(n-1)} + v^n \sum_{k=1}^{m} p_{ik}^{(n-1)} \left(\psi_k(n) + \sum_{j=1}^{m} p_{kj} \gamma_{kj}(n) \right), \qquad [3.143]$$

$$V_i^{(n)}(s) = V_i^{(n-1)}(s) + v^n \sum_{k=1}^{m} p_{ik}^{(n-1)}(s) \left(\psi_k(s+n) + \sum_{j=1}^{m} p_{kj}(s+n) \gamma_{kj}(s+n) \right). \qquad [3.144]$$

Relations [3.143] and [3.144] outline the presence in the RWMPV of the lump sums given or taken in the instant of the state change.

3.5.2.1.6. Sixth model

Now, we will present the formulas in the same previous assumptions but taking into account t.he variability of interest rates.

More precisely, we assume that:

1) rewards are variable in time;

2) rewards are given for permanence in the state and at a given transition;

3) rewards are paid at the end of the period;

4) the interest rates are time-dependent.

In this case, we obtain the following homogeneous and non-homogeneous formulas:

$$V_i^{(n)} = V_i^{(n-1)} + v(n) \sum_{k=1}^{m} p_{ik}^{(n-1)} \left(\psi_k(n) + \sum_{j=1}^{m} p_{kj} \gamma_{kj}(n) \right), \qquad [3.145]$$

$$V_i^{(n)}(s) = V_i^{(n-1)}(s) \\ +v(s,s+n)\sum_{k=1}^{m} p_{ik}^{(n-1)}(s)\left(\psi_k(s+n)+\sum_{j=1}^{m} p_{kj}(s+n)\gamma_{kj}(s+n)\right). \qquad [3.146]$$

3.5.2.2. *Due cases*

In the discounted case, the difference between the due and immediate cases is of great relevance. However, in our opinion, it is enough to present some formulas for the due case because of their resemblance to the immediate case.

3.5.2.2.1. First model

Here, the basic assumptions are:

1) rewards are fixed in time;

2) rewards are given only for permanence in the state;

3) rewards are paid at the beginning of the period;

4) the interest rate r is fixed.

In this case, $\ddot{V}_i^{(n)}$ represents the RWMPV given that at time 0 the system was in the state i and the rewards were paid at the beginning of the period. Under these hypotheses, for the homogeneous case, the following formulas are obtained:

$$\ddot{V}_i^{(1)} = \psi_i, \qquad [3.147]$$

$$\ddot{V}_i^{(2)} = \psi_i + (1+r)^{-1}\sum_{k=1}^{m} p_{ik}\psi_k = \ddot{V}_i^{(1)} + (1+r)^{-1}\sum_{k=1}^{m} p_{ik}^{(1)}\psi_k, \qquad [3.148]$$

$$\ddot{V}_i^{(n)} = \ddot{V}_i^{(n-1)} + (1+r)^{-n+1}\sum_{k=1}^{m} p_{ik}^{(n-1)}\psi_k, \qquad [3.149]$$

and in the non-homogeneous case:

$$\ddot{V}_i^{(n)}(s) = \ddot{V}_i^{(n-1)}(s) + (1+r)^{-n+1}\sum_{k=1}^{m} p_{ik}^{(n-1)}(s)\psi_k. \qquad [3.150]$$

Relations [3.149] and [3.150] give the RWMPV in the simplest due cases (fixed interest rate and fixed rewards). Also, in these cases, the recursive nature of the formulas is preserved.

3.5.2.2.2. Second model

Here, basic assumptions are:

1) rewards are variable in time;

2) rewards are given only for permanence in the state;

3) rewards are paid at the beginning of the period;

4) the interest rates are time-dependent.

For the homogeneous case, related formulas are:

$$\ddot{V}_i^{(1)} = \psi_i(0) \ , \qquad [3.151]$$

$$\ddot{V}_i^{(2)} = \ddot{V}_i^{(1)} + v(1)\sum_{k=1}^{m} p_{ik}^{(1)}\psi_k(1), \qquad [3.152]$$

$$\ddot{V}_i^{(n)} = \ddot{V}_i^{(n-1)} + v(n-1)\sum_{k=1}^{m} p_{ik}^{(n-1)}\psi_k(n-1). \qquad [3.153]$$

For the non-homogeneous case, we obtain in general:

$$\ddot{V}_i^{(n)}(s) = \ddot{V}_i^{(n-1)}(s) + v(s,s+n-1)\sum_{k=1}^{m} p_{ik}^{(n-1)}(s)\psi_k(s+n-1). \qquad [3.154]$$

The discount factors and indices of the reward are given by the moment of payment and are therefore one period before, i.e. the payment time. Also, in this case, RWMPV can be obtained recursively.

3.5.2.2.3. Third model

Now, we introduce the γ rewards. We make the following assumptions:

1) rewards are variable in time;

2) rewards are given for permanence in the state and at a given transition;

3) rewards are paid at the beginning of the period;

4) the interest rate r is fixed.

Under these hypotheses, the general recursive formulas for homogeneous and non-homogeneous cases are the following:

$$\ddot{V}_i^{(n)} = \ddot{V}_i^{(n-1)} + v^{n-1}\sum_{k=1}^{m} p_{ik}^{(n-1)}\left(\psi_k(n-1) + v\sum_{j=1}^{m} p_{kj}\gamma_{kj}(n)\right), \qquad [3.155]$$

$$\begin{aligned}&\ddot{V}_i^{(n)}(s) = \ddot{V}_i^{(n-1)}(s) + \\ &v^{n-1}\sum_{k=1}^{m} p_{ik}^{(n-1)}(s)\left(\psi_k(s+n-1) + v\sum_{j=1}^{m} p_{kj}(s+n)\gamma_{kj}(s+n)\right).\end{aligned} \qquad [3.156]$$

In the RWMPV, the lump sums given or taken in the instant of the state changes are present. This is why the recursivity holds.

3.5.2.2.4. Fourth model

The next case, which only makes sense in a non-homogeneous environment, takes into account the non-homogeneity in interest rate and rewards. In this case, the following hypotheses hold:

1) rewards are variable because of the starting time and the time spent after;

2) rewards are given also at the transitions;

3) rewards are paid at the beginning of the period;

4) the interest rate is variable because of the starting time and the time spent after.

We obtain:

$$\begin{aligned}\ddot{V}_i^{(n)} &= \ddot{V}_i^{(n-1)} + \dot{v}(s,s+n-1)\sum_{k=1}^{m} p_{ik}^{(n-2)}(s)\sum_{j=1}^{m} p_{kj}(s+n-1)\psi_{kj}(s,s+n-1) \\ &+ \dot{v}(s,s+n)\sum_{k=1}^{m} p_{ik}^{(n-1)}(s)\sum_{j=1}^{m} p_{kj}(s+n)\gamma_{kj}(s,s+n).\end{aligned} \qquad [3.157]$$

3.6. General algorithms for the DTMRWP

In this section, we will briefly show two algorithms useful for the solution, respectively, of homogeneous and non-homogeneous cases. Each of them can be used for all the different homogeneous and non-homogeneous cases.

This fact means that with a unique algorithm, it is possible to treat all the possible different homogeneous cases and by means of another algorithm all non-homogeneous cases can be approached.

The algorithms will work for a time horizon of T periods and will be able to tackle any combinations of the cases considered above.

The algorithms are constructed following structured programming rules [DAH 72] and we will only summarize the steps:

3.6.1. *Homogeneous MRWP*

Input – reads the selectors of the examined case, number of states and the number of periods, the permanence and transition rewards, the fixed or variable interest rate, and the transition matrix **P**.

Compute – matrix for all kinds of permanence rewards.

Compute – matrix for all kinds of transition rewards.

Compute – vector of discount factors (in the non-discounted case is the *sum* vector).

Compute – reward matrix.

The *sum* vector is the vector whose elements are all equal to 1.

3.6.2. *Non-homogeneous MRWP*

Input – reads the selectors of the examined case, the number of states and the number of periods, the permanence and transition rewards, the fixed or variable interest rate, and the transition matrix **P**(s).

Compute – matrix for all kinds of permanence rewards.

Compute – matrix for all kinds of transition rewards.

Compute – matrix of discount factors (in the non-discounted case is the **1** matrix).

Compute – reward matrix.

More details are available from to the authors upon request.

4

Homogeneous and Non-homogeneous Semi-Markov Models

4.1. Continuous time semi-Markov processes

In this chapter, the continuous time homogeneous semi-Markov process (CTHSMP) and continuous time non-homogeneous semi-Markov process (CTNHSMP) will be described following the Semi-Markov process (SMP) notation given by Janssen and Manca [JAN 06].

As in Chapter 3, let us consider a *physical* or *economic system* called S with m possible states forming the set

$$I = \{1, \ldots, m\}. \qquad [4.1]$$

At time 0, system S starts from an initial state represented by the r.v. J_0, stays a non-negative random length of time X_1 in this state, and then goes into another state J_1 for a non-negative length of time X_2 before going into J_2, etc.

So we have a two-dimensional (2D) stochastic process in discrete time called a *positive (J-X) process* or simply *(J-X) process*

$$(J - X) = ((J_n, X_n), n \geq 0) \qquad [4.2]$$

assuming

$$X_0 = 0, \; a.s. \qquad [4.3]$$

where the sequence $(J_n, n \geq 0)$ gives the successive *states* of S in time and the sequence $(X_n, n \geq 0)$ gives the successive *sojourn times* spent in state J_{n-1}.

Times at which transitions occur are given by the sequence $(T_n, n \geq 0)$ where:

$$T_0 = 0, T_1 = X_1, \ldots, T_n = \sum_{r=1}^{n} X_r \qquad [4.4]$$

and so

$$X_n = T_n - T_{n-1}, n \geq 1 \qquad [4.5]$$

DEFINITION 4.1.– *The two-dimensional process (J,T) = $((J_n, T_n), n \geq 0)$ where T_n is given by relation [4.4] is called a Markov renewal sequence or Markov renewal process.*

Çinlar [ÇIN 69] also gives the term *Markov additive process.*

On the complete probability space $(\Omega, \Im, P)$, the stochastic dynamic evolution of the considered (*J-X*) process will be called a *homogeneous semi-Markov process* (in short CTHSMP) determined by the following assumptions:

$P(X_0=0))=1$, a.s.,

$$P(J_0=i)=p_i,\ i=1,\ldots,m \text{ with } \sum_{i=1}^{m} p_i = 1, \qquad [4.6]$$

and moreover for all $n>0, j=1,\ldots,m,$ we assume the semi-Markov condition

$$P(J_n = j, X_n \leq x \mid (J_k, X_k), k = 0, \ldots, n-1) = Q_{J_{n-1},j}(x), a.s. \qquad [4.7]$$

where any function Q_{ij} $(i,j=1,\ldots,m)$ is a non-decreasing real function on $\mathbb{R}^+$ such that if

$$p_{ij} = \lim_{x \to +\infty} Q_{ij}(x),\ i,j \in I, \qquad [4.8]$$

then:

$$\sum_{j=1}^{m} p_{ij} = 1,\ i \in I. \qquad [4.9]$$

Equivalently, using relations [4.4] and [4.5], the basic condition [4.7] can also be written as follows

$$\begin{aligned} P(T_n - T_{n-1} \le x, J_n = j | (J_k, T_k), k = 0,...,n-1, J_{n-1} = i) \\ = P(T_n - T_{n-1} \le x, J_n = j | J_{n-1} = i)(= Q_{ij}(x)). \end{aligned} \quad [4.10]$$

With matrix notation, we will write:

$$\mathbf{Q}(x) = [Q_{ij}(x)],\ \mathbf{P} = [p_{ij}](= \mathbf{Q}(\infty)),\ \mathbf{p} = (p_1,..., p_m). \quad [4.11]$$

To define a *non-homogeneous semi-Markov process* (in short CTHSMP), we replace condition [4.10] by the following condition

$$\begin{aligned} P(T_n \le t, J_n = j | (J_k, T_k), k = 0,...,n-2, J_{n-1} = i, T_{n-1} = s) \\ = P(T_n \le t, J_{n-1} = i, T_{n-1} = s)(= Q_{ij}(s,t). \end{aligned} \quad [4.12]$$

Here, we have

$$\mathbf{Q}(s,t) = [Q_{ij}(s,t)],\ \mathbf{P}(s) = [p_{ij}(s)](= \mathbf{Q}(s,\infty)),\ \mathbf{p} = (p_1,..., p_m) \quad [4.13]$$

where $\mathbf{P}(s) = [p_{ij}(s)]$ is a Markov transition matrix for all positive t.

It is also necessary to introduce the probability that the process will leave state i in a time t:

$$\begin{aligned} H_i(t) = P[T_{n+1} - T_n \le t | J_n = i], \\ H_i(s,t) = P[T_{n+1} - T_n \le t | J_n = i, T_n = s]. \end{aligned} \quad [4.14]$$

Clearly, these functions are distribution functions (d.f.) and

$$\begin{aligned} H_i(t) = \sum_{j=1}^{m} Q_{ij}(t), \\ H_i(s,t) = \sum_{j=1}^{m} Q_{ij}(s,t). \end{aligned} \quad [4.15]$$

Now the d.f. of the waiting time in each state i can be defined, given that the state successively occupied is known:

$F_{ij}(t)$=P[T_{n+1}-T_n<t | $J_n = i$, $J_{n+1} = j$].

$F_{ij}(s,t)$=P[T_{n+1}<t | $J_n = i$, $J_{n+1} = j, T_n$=s].

The difference between continuous Markov and SMPs is mainly in the form of these last d.f. Indeed, in the Markov case, these d.f. can only be negative exponential in continuous time and geometric in discrete time whereas in the semi-Markov case, they could be of any type.

The related probabilities can be obtained by means of the following formula:

$$F_{ij}(t) = \begin{cases} \dfrac{Q_{ij}(t)}{p_{ij}} & \text{if } \; p_{ij} \neq 0, \\ U_1(t) & \text{if } \; p_{ij} = 0 \end{cases} \qquad [4.16]$$

$$F_{ij}(s,t) = \begin{cases} \dfrac{Q_{ij}(s,t)}{p_{ij}(s)} & \text{if } \; p_{ij}(s) \neq 0, \\ U_1(s,t) & \text{if } \; p_{ij}(s) = 0 \end{cases}, \qquad [4.17]$$

where:

$$U_1(t) = 1. \text{ and } U_1(s,t) = \begin{cases} 0 & \text{if } \; s > t \\ 1 & \text{if } \; s \leq t \end{cases} \qquad [4.18]$$

REMARK 4.1.–

(1) From relations [4.16] and [4.17], we can also express kernel **Q** as a function of F_{ij}, i,j=1,...,m:

$$\begin{aligned} &Q_{ij}(x) = p_{ij}F_{ij}(x), i, j \in I, x \in \mathbb{R}^+, \\ &Q_{ij}(s,t) = p_{ij}(s)F_{ij}(s,t), i, j \in I, s, t \in \mathbb{R}^+ (s < t). \end{aligned} \qquad [4.19]$$

So, every SMC can also be characterized by the triple (**p**,**P**,**F**) instead of the couple (**p**,**Q**) where the $m \times m$ matrix **F** is defined as $\mathbf{F} = \left[F_{ij}\right]$, and where the functions F_{ij}, $i, j = 1, \ldots, m$ are d.f. on support $\mathbb{R}^+$.

(2) We can also introduce the *means* related to these conditional and unconditional d.f.

When they exist we will note:

$$\begin{aligned}\beta_{ij} &= \int_R x dF_{ij}(x), \left(\beta_{ij}(s) = \int_R x dF_{ij}(s,x) \right) i,j = 1,...,m, \\ \eta_i &= \int_R x dH_i(x), \left(\eta_i(s) = \int_R x dH_i(s,x) \right), i = 1,...,m\end{aligned} \quad [4.20]$$

and relations [4.15] and [4.20] lead to the relation:

$$\eta_i = \sum_{j=1}^{m} p_{ij}\beta_{ij} \left(\eta_i(s) = \sum_{j=1}^{m} p_{ij}(s)\beta_{ij}(s) \right). \quad [4.21]$$

The quantities $\beta_{ij}(\beta_{ij}(s))$, $i,j = 1,...,m$ and $\eta_i(\eta_i(s)), I = 1,...,m$ are respectively called the *conditional* and *unconditional means* of the sojourn times.

4.2. The embedded Markov chain

We will start by studying the marginal stochastic processes $(J_n, n \geq 0)$, $(X_n, n \geq 0)$ called the *J-process* and the *X-process,* respectively.

The J-process

From properties of the conditional expectation, the process $(J_n, n \geq 0)$ satisfies the following property:

$$P(J_n = j | (J_k, T_k), k = 0,...,n-2, T_{n-1} = s) = Q_{J_{n-1}j}(+\infty)(= Q_{J_{n-1}j}(s,+\infty). \quad [4.22]$$

Using the smoothing property of conditional expectation, we get

$$\begin{aligned}&P(J_n = j | (J_k), k = 0,...,n-1) = E(Q_{J_{n-1}j}(+\infty) | (J_k), k = 0,...,n-1) \\ &(= E(Q_{J_{n-1}j}(s,+\infty) | (J_k), k = 0,...,n-1))\end{aligned} \quad [4.23]$$

and so

$$\begin{aligned}&P(J_n = j | (J_k), k = 0,...,n-1) = p_{J_{n-1}j}, \\ &P(J_n = j | (J_k), k = 0,...,n-1, T_{n-1} = s) = p_{J_{n-1}j}(s).\end{aligned} \quad [4.24]$$

Since relation [4.24] implies that matrix **P** is a Markov matrix, we have thus proved the following result.

PROPOSITION 4.1.– The J-process is a homogeneous (respectively, non-homogeneous) Markov chain with **P** (respectively, **P**(s)) as its transition matrix.

That is the reason why this *J*-process is called the *embedded Markov chain* of the considered SMC in which the r.v. J_n represents the state of the system *S* just after the *n*th transition.

Let us remark that for the non-homogeneous case, this chain depends on T_n and so it is better to speak of a *multiple embedded Markov chain.*

4.3. The counting processes and the associated semi-Markov process

We will now define the *counting processes* associated with any Markov renewal process (MRP).

For any fixed time *t*, the r.v. $N(t)$ represents the *total number of jumps or transitions of the* (J,X) *process on* $(0,t]$, including possible transitions from any state toward itself (virtual transitions), assuming transitions are observable.

We have:

$$N(t) > t \Leftrightarrow T_n \le t . \qquad [4.25]$$

However, here, we can be more precise and only count the total number of passages in a fixed state i always in $(0,t]$ represented by the r.v. $N_i(t)$.

Clearly, we can write:

$$N(t) = \sum_{i=1}^{m} N_i(t), t \ge 0 . \qquad [4.26]$$

DEFINITION 4.2.– *With each Markov renewal process, the following m+1 stochastic processes are associated respectively with values in* $\mathbb{N}$:

(1) the *N*-process $(N(t), t \ge 0)$;

(2) the N_i-process $(N_i(t), t \ge 0)$, $i=1,\ldots,m$,

respectively called the associated total counting process and the associated partial counting processes with of course:

$$N(0)=0, N_i(0)=0, i=1,...,m. \qquad [4.27]$$

It is now easy to introduce the notion of a *SMP* by considering at time t, the state entered at the last transition before or at t, that is, $J_{N(t)}$.

DEFINITION 4.3.– *With each Markov renewal process, we associate the following stochastic Z-process with values in I:*

$$Z=(Z(t),\ t. \geq 0\ .), \qquad [4.28]$$

with:

$$Z(t)=J_{N(t)}. \qquad [4.29]$$

This process will be called the associated SMP or simply the SMP of kernel $\mathbf{Q}$.

The transition probabilities of the SMP Z are defined in the following way:

$$\begin{aligned} \phi_{ij}(t) &= P[\ Z_t = j \mid Z_0 = i\ ], \\ \phi_{ij}(s,t) &= P[\ Z_t = j \mid Z_s = i\ ]. \end{aligned} \qquad [4.30]$$

They are obtained solving the following evolution equations:

$$\phi_{ij}(t) = \delta_{ij}(1-H_i(t)) + \sum_{\beta=1}^{m} \int_0^t \dot{Q}_{i\beta}(\vartheta)\phi_{\beta j}(t-\vartheta)d\vartheta \qquad [4.31]$$

$$\phi_{ij}(s,t) = \delta_{ij}(1-H_i(s,t)) + \sum_{\beta=1}^{m} \int_s^t \dot{Q}_{i\beta}(s,\vartheta)\phi_{\beta j}(\vartheta,t)d\vartheta \qquad [4.32]$$

where δ_{ij} represents the Kronecker symbol, $\dot{Q}_{ik}$ is the derivative of the Q_{ik} and could be seen as the probability intensity of changing the state i just at time ϑ.

Both [4.28] and [4.29] can be obtained by means of a simple probabilistic argument using the regenerative property of the Markov renewal process (see Janssen & Manca [JAN 06, JAN 07]).

With the aim of clarification, the meaning of the parts of [4.31] is given as:

$$\delta_{ij}(1 - H_i(s,t))$$

represents the probability of remaining in the state i without any transition from the time s up to time t, and it only makes sense if $i = j$; this is the reason for the Kronecker delta.

$$\sum_{\beta=1}^{m}\int_{s}^{t}\dot{Q}_{i\beta}(s,\vartheta)\phi_{\beta j}(\vartheta,t)d\vartheta$$

represents the probabilities of all the possible trajectories that can be done to go from state i to state j given that the system arrived in the state i at time s and it is in the state j at time t.

In Figure 4.1, trajectory of a SMP is shown.

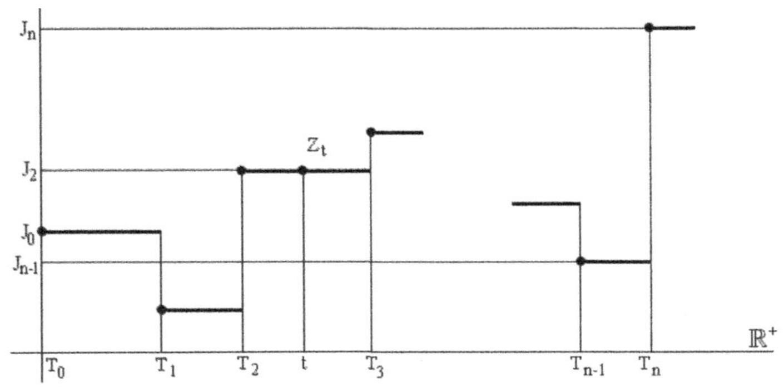

Figure 4.1. *A trajectory of a SMP*

4.4. Initial backward recurrence times

Now the initial backward recurrence time will be introduced. To explain the meaning of backward time, we show Figures 4.2 and 4.3 in which homogeneous and non-homogeneous cases are shown.

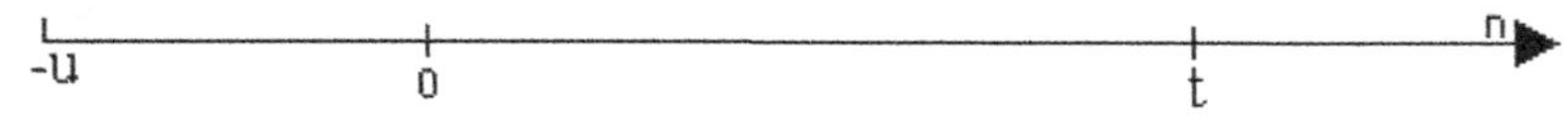

Figure 4.2. *Homogeneous SMP backward time*

Figure 4.3. *Non-homogeneous SMP backward time*

As in the "simple" SMP we follow the system from time s up to time t, but, this time, we consider for example in a non-homogeneous environment, that the system entered the state i at time s-u and that the system does not move from i for a period of time u that is the recurrence backward time.

Now, in order to take into account the backward time, it is necessary to condition the system to remain for a time u inside the state i.

The last two relations present the backward evolution equations of the homogeneous and non-homogeneous SMP. Under the backward assumptions, relations [4.31] and [4.32] become respectively:

$$\begin{aligned}\phi_{ij}(u,t) &= \delta_{ij}\frac{(1-H_i(t+u))}{(1-H_i(u))}+\sum_{\beta=1}^{m}\int_u^t \frac{\dot{Q}_{i\beta}(\vartheta+u)}{(1-H_i(u))}\phi_{\beta j}(0,t-\vartheta)d\vartheta,\\ \phi_{ij}(u,s,t) &= \delta_{ij}\frac{(1-H_i(s-u,t))}{(1-H_i(s-u,s))}+\sum_{\beta=1}^{m}\int_s^t \frac{\dot{Q}_{i\beta}(s-u,\vartheta)}{(1-H_i(s-u,s))}\phi_{\beta j}(0,\vartheta,t)d\vartheta.\end{aligned} \quad [4.33]$$

where

$$\phi_{ij}(s,t)=P\left[X_{n+1}=j\middle|X_n=i,T_{n+1}>t,T_n=s\right],i,j\in I,s,t\in\mathbb{R}^+,s<t. \quad [4.34]$$

The two relations are obtained supposing the probability that the system will remain for a time u in state i and that the system is at time 0 in state i, in the homogeneous case. In this case, it is supposed that the system arrived at time u before 0 in the state i. In the non-homogeneous case, it is supposed that the system arrived at time $s-u$ in the state i and does not move from this state up to the time s.

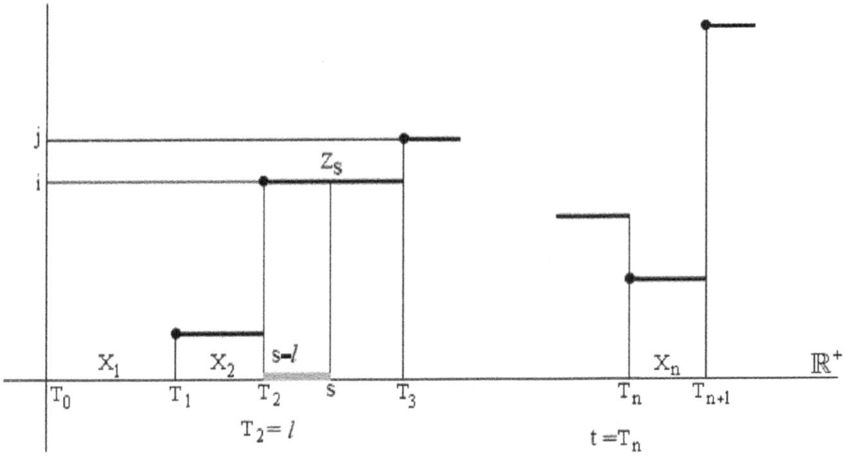

Figure 4.4. *Non-homogeneous semi-Markov backward trajectory*

4.5. Particular cases of MRP

We will devote this section to particular cases of MRP having the advantage to lead to some explicit results.

4.5.1. *Renewal processes and Markov chains*

For the sake of completeness, let us first state that with $m = 1$, that is, that the observed system has only one possible state, the kernel **Q (Q(s))** has only one element, say the d.f. **F (F(s))**, and the process $(X_n, n > 0)$ is a *renewal process*.

Second, to obtain *Markov chains* studied in Chapter 4, it suffices to choose for matrix **F** the following special degenerating case:

$$F_{ij} = U_1, (F_{ij}(s,t) = U_s(t), < t) \forall i, j \in I \qquad [4.35]$$

and of course an arbitrary Markov matrix **P (P(s))**.

This means that all r.v. X_n have a.s. the value 1, and so the single random component is the (J_n) process, which is a homogeneous (repectively, non-homogeneous) MC of transition matrix **P (P(s))**.

4.5.2. *MRP of zero-order (PYKE (1962))*

There are two types of such processes.

4.5.2.1. *First type of zero-order MRP*

This type is defined by the following semi-Markov kernel

$$\mathbf{Q} = [p_i F_i]([p_i(s) F_i(s)]) \qquad [4.36]$$

so that:

$$p_{ij} = p_i (p_{ij}(s) = p_i(s)), F_{ij} = F_i (F_{ij}(s,t) = F_i(s,t), s < t), j \in I. \qquad [4.37]$$

Naturally, we assume that for every i belonging to I, p_i is strictly positive.

In the homogeneous case, we discover that the r.v. $J_n, n \geq 0$ are independent and identically distributed and moreover that the conditional inter-arrival distributions do not depend on the state to be reached, so that, by relation [4.15],

$$H_i = F_i, i \in I. \qquad [4.38]$$

Moreover, since:

$$P(X_n \leq x|(J_k, X_k), k \leq n-1, J_n) = F_{J_{n-1}}(x), \tag{4.39}$$

we obtain:

$$P(X_n \leq x|(X_k), k \leq n-1) = \sum_{j=1}^{m} p_j F_j(x). \tag{4.40}$$

Introducing the d.f. F defined as

$$F = \sum_{j=1}^{m} p_j F_j, \tag{4.41}$$

the preceding equality shows that, for a homogeneous MRP of zero-order of the first type, the sequence $(X_n, n \geq 1)$ is a renewal process characterized by the d.f. F.

4.5.2.2. *Second type of zero-order MRP*

This type is defined by the following semi-Markov kernel

$$\mathbf{Q} = [p_i F_j](\mathbf{Q}(s) = [p_i(s) F_j(s)], \tag{4.42}$$

so that:

$$p_{ij} = p_i (p_{ij}(s) = p_i(s), F_{ij}(s,t) = F_j(t) s < t), \; i, j \in I. \tag{4.43}$$

Here too, we suppose that for every i belonging to I, p_i is strictly positive.

Once again, in the homogeneous case, the r.v. $J_n, n \geq 0$ are independent and identically distributed and moreover the conditional inter-arrival distributions do not depend on the state to be *left*, so that, by relation [4.15]

$$H_i = \sum_{j=1}^{m} p_j F_j (= F), i \in I. \tag{4.44}$$

Moreover, since:

$$P(X_n \leq x|(J_k, X_k), k \leq n-1, J_n) = F_{J_n}(x), \tag{4.45}$$

we obtain

$$P(X_n \leq x|(X_k), k \leq n-1) = \sum_{j=1}^{m} p_j F_j(x) = F(x). \qquad [4.46]$$

The preceding equality shows that, for a homogeneous MRP of zero-order of the second type, the sequence $(X_n, n \geq 1)$ is a renewal process characterized by the d.f. F as in the first type.

The basic reason for these similar results is that these two types of MRP are the *reverses* (timewise) of each other.

4.5.3. *Continuous Markov processes*

These processes are defined by the following particular semi-Markov kernel

$$\mathbf{Q}(x) = \left[p_{ij}\left(1-e^{-\lambda_i x}\right)\right], x \geq 0\left(\mathbf{Q}(s,t) = \left[p_{ij}(s)\left(1-e^{-\lambda_i (s)(t-s)}\right)\right]\right), \qquad [4.47]$$

where $\mathbf{P} = \left[p_{ij}\right]\left(\left[p_{ij}(s)\right]\right)$ is a stochastic matrix and where parameters $\lambda_i, (\lambda_i(t-s), s < t), i \in I$ are strictly positive.

The homogeneous standard case corresponds to that in which $p_{ii} = 0, i \in I$ (see [CHU 60]).

From relation [4.47], we obtain:

$$F_{ij}(x) = 1 - e^{-\lambda_i x}. \qquad [4.48]$$

Thus, the d.f. of sojourn time in state i has an exponential distribution depending uniquely upon the occupied state i, such that both the excess and age processes also have the same distribution.

For $m = 1$, we obtain the usual Poisson process of parameter $\lambda(\lambda(t-s), s < t)$.

4.6. Examples

Semi-Markov theory is one of the most productive subjects of stochastic processes to generate applications in real-life problems, particularly in the following fields: economics, manpower models, insurance, finance (more recently), reliability, simulation, queuing, branching processes, medicine (including survival data), social sciences, language modeling, seismic risk analysis, biology, computer science, chromatography and fluid mechanics.

Important results in such fields may be found in Janssen [JAN 86], Janssen & Limnios [JAN 99a], Devolder *et al.* [DEV 12], Janssen *et al.* [JAN 13], Janssen & Manca [JAN 06, JAN 07],

Let us give three examples in the fields of insurance and reliability.

EXAMPLE 4.1.– The claim process in insurance

Let us consider an insurance company covering m types of risks or having m different types of customers for the same risk forming the set $I=\{1,...,m\}$.

For example, in automobile insurance, we can distinguish three types of driver: *good*, *average* and *bad* and so I is a space consisting of three states: 1 for good, 2 for average and 3 for bad.

Now, let $(X_n, n \geq 1)$ represent the sequence of successive observed *claim amounts*, $(Y_n, n \geq 1)$ the sequence of inter-arrival times between two successive claims and $(J_n, n \geq 1)$ successive *types of observed risks*.

In the traditional model of risk theory called the "Cramer Lundberg model" [CRA 55, LUD 09], it is assumed with that there is only one type of risk and the claim arrival process is a *Poisson process* parameter λ. Later, Andersen [AND 67] extended this model to an arbitrary *renewal process* and moreover in these two traditional models, the process of claim amounts is a renewal process independent of the claim arrival process.

The consideration of an SMC for the two-dimensional processes $((J_n, X_n), n \geq 0)$ and/or $((J_n, Y_n), n \geq 0)$ provides the possibility to introduce a certain dependence between the successive claim amounts. This model was first developed by Janssen [JAN 69, JAN 70, JAN 77] along the lines of Miller's work (1962) [MIL 62] and since then has led to many extensions; see for example Asmussen [ASM 00].

EXAMPLE 4.2.– Occupational illness insurance

This problem is related to occupational illness insurance with the possibility of leading to partial or permanent disability. In this case, the amount of the incapacitation allowance depends on the degree of disability recognized in the policyholder by the occupational health doctor, in general on an annual basis, because this degree is a function of an occupational illness which can take its course.

Considering as in the example in section 11.2.2 of [JAN 07] this invalidity degree as a stochastic process $(J_n, n \geq 0)$, where J_n represents the value of this degree when the illness really takes its course, we must then introduce the r.v. X_n representing the time between two successive transitions from J_{n-1} to J_n.

In practice, these transitions can be observed with periodic medical inspections. The assumption that the *J-X* process is an SMC extends the Markov model of Chapter 11 [JAN 07] and is fully treated in Janssen & Manca [JAN 06, JAN 07].

EXAMPLE 4.3.– Reliability

There are many semi-Markov models in reliability theory; see for example Osaki [OSA 85] and more recently Limnios & Oprisan [LIM 01, LIM 03].

Let us consider a *reliability system S* that can be at any time t in one of the m states of $I=\{1,\dots,m\}$.

The stochastic process of the successive states of S is represented by $S=(S_t, t \geq 0)$.

The state space I is partitioned into two sets U and D so that

$$I = U \cup D,\ U \cap D = \varnothing,\ U \neq \varnothing,\ D \neq \varnothing. \qquad [4.49]$$

The interpretation of these two sets is the following: the subset U contains all "good" states, in which the system is working and the subset D of all "bad" states, in which the system is not working well or has failed.

The indicators used in reliability theory are the following:

(1) the *reliability function R* gives the probability that the system was always working from time 0 to time t:

$$R(t) = P\left(S_u \in U, \forall u \in [0,t]\right), \qquad [4.50]$$

(2) the *pointwise availability function A* gives the probability that the system is working at time t whatever happens on $(0,t)$:

$$A(t) = P\left(S_t \in U\right), \qquad [4.51]$$

(3) the *maintainability function* M gives the probability that the system, being in D on $[0,t)$, will leave set D at time t:

$$M(t)=P\left(S_u \in D, u\in[0,t), S_t \in U\right). \qquad [4.52]$$

4.7. Discrete time homogeneous and non-homogeneous semi-Markov processes

In this section, we adapt the general presentation given above when the time is given on a discrete time scale, that is $T = \{0,1,\dots,n,\dots\}$. In particular, this is very useful for numerical computation. We now introduce the following definitions:

$$b_{ij}(t)=P\left[J_{n+1}=j, T_{n+1}-T_n=t \mid J_n=i\right]=\begin{cases} Q_{ij}(t)-Q_{ij}(t-1) & \text{if } t>0 \\ 0 & \text{if } t=0 \end{cases},$$

$$b_{ij}(s,t)=P\left[J_{n+1}=j, T_{n+1}=t \mid J_n=i, T_n=s\right] \qquad [4.53]$$

$$=\begin{cases} Q_{ij}(s,t)-Q_{ij}(s,t-1) & \text{if } t>s \\ 0 & \text{if } t=s \end{cases}$$

where

$$H_i(t)=\sum_{j=1}^{m} Q_{ij}(t),$$

$$H_i(s,t)=\sum_{j=1}^{m} Q_{ij}(s,t). \qquad [4.54]$$

For the transition probabilities of the SMP

$$\phi_{ij}(t)=\mathrm{P}\left[Z(t)=j \mid Z(0)=i\right].$$

$$\phi_{ij}(s,t)=\mathrm{P}\left[Z(t)=j \mid Z(s)=i\right].$$

the evolution equations are now given by

$$\phi_{ij}(t)=\delta_{ij}(1-H_i(t))+\sum_{\beta=1}^{m}\sum_{\vartheta=1}^{t} b_{i\beta}(\vartheta)\phi_{\beta j}(t-\vartheta) \qquad [4.55]$$

$$\phi_{ij}(s,t)=\delta_{ij}(1-H_i(s,t))+\sum_{\beta=1}^{m}\sum_{\vartheta=s+1}^{t} b_{i\beta}(s,\vartheta)\phi_{\beta j}(\vartheta,t) \qquad [4.56]$$

The first terms of formulas [4.55] and [4.56]

$$\delta_{ij}(1 - H_i(t)) \tag{4.57}$$

$$\delta_{ij}(1 - H_i(s,t)) \tag{4.58}$$

give the probabilities that the system does not have transitions up to the time t given that it was in the state i at time 0 in the homogeneous case and at time s in the non-homogeneous.

Functions $b_{i\beta}(\vartheta)$ and $b_{i\beta}(s,\vartheta)$ represent probabilities that the system enters in the state β just at time ϑ given that it entered in the state i respectively at time 0 and at time s. After the transition, the system will go to the state j following one of the possible trajectories that go from the state β to the state j within the time t. In the credit risk environment, it means that up to the time ϑ the rating company does not give any other evaluation of the firm; at time ϑ the rating company gave the new rating β to the evaluated firm. After this, the rating will arrive to the state j within the time t following one of the possible rating trajectories.

REMARK 4.2.–

(1) Let us recall that any SMP is characterized by its kernel $\mathbf{Q}(s)$ or $\mathbf{Q}(s,t)$ and eventually an initial condition on $Z(0)(= J_0)$ given by the vector $\mathbf{a}$

$$\begin{aligned} &\mathbf{a} = (a_1, \dots, a_m) \\ &a_i = P(J_0 = i), i = 1, \dots, m, \\ &\sum_{i=1}^{m} a_i = 1. \end{aligned} \tag{4.59}$$

Indeed, if we know kernels $\mathbf{Q}(s)$ or $\mathbf{Q}(s,t)$, non-homogeneous case by the relations [4.54], it is possible to compute the d.f. $H_i(t)$ and the $H_i(s,t)$. Furthermore, it is also possible to compute the functions $b_{ij}(t)$ and the $b_{ij}(s,t)$ by [4.53]. Consequently, it is also possible to solve the linear equations [4.55] and [4.56] as described in Corradi *et al.* [COR 04] and Janssen & Manca [JAN 01].

2) From real data, functions $Q_{ij}(t)$ and $Q_{ij}(s,t)$ can be more easily obtained by relation [4.19] . Indeed, waiting time d.f. and the embedded Markov chain matrices can be obtained directly by the data. So, having the real data, it is possible to estimate quite easily semi-Markov kernels in both homogeneous and non-homogeneous cases. Of course, we need much more data for the non-homogeneous case.

4.8. Semi-Markov backward processes in discrete time

4.8.1. *Definition in the homogeneous case*

DEFINITION 4.4.– *Let* $B(t) = t - T_{N(t)}$ *be the backward recurrence time process in a semi-Markov environment is defined by:* $B(t) = t - T_{N(t)}$.

Then we define the following probability:

$$^{b}\phi_{ij}(l;t) = \mathrm{P}\left[Z(t) = j \mid Z(0) = i, B(0) = l\right], \qquad [4.60]$$

where [4.60] represents the semi-Markov transition probabilities with initial backward recurrence time as shown in Figure 4.5.

In Figure 4.3, a trajectory of an HSMP with initial backward recurrence time is reported. In a homogeneous environment, the system starts from time 0. We have that $N(0) = n$, because we start to follow our system after the n-th transition. The starting backward is $B(0) = l$ then $T_n = -l$ represents, in function of homogeneous hypothesis, the time of the n-th transition and J_n the related state. The time t represents the duration from 0. $J_{h-1} = j$ the state of the $h-1$-th transition, T_{h-1} the time of arrival in the state *j* and $N(t) = h-1,\ h-1 > n$.

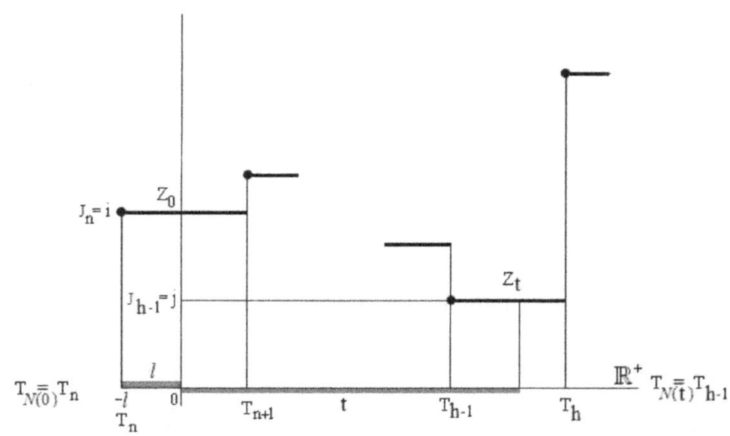

Figure 4.5. *HSMP with backward time trajectory*

To present the evolution equations of probabilities [4.60], we introduce the following notation:

$$D_i(l;t) = \frac{1 - H_i(l+t)}{1 - H_i(l)} \qquad [4.61]$$

which represents the probability of having no transition from state i between times $-l$ and t given that no transition occured from state i between times $-l$ and 0.

Moreover

$$b_{ij}(l;t) = \frac{b_{ij}(l+t)}{1 - H_i(l)} \qquad [4.62]$$

represents the probability of making the next transition from state i to state j just at time t given that the system does not make transitions from state i between times $-l$ and 0.

So, we obtain the following evolution equations

$$^b\phi_{ij}(l;t) = D_i(l;t) + \sum_{\beta \in E} \sum_{\vartheta = s+1}^{t} b_{i\beta}(l;\vartheta)\phi_{\beta j}(t-\vartheta), \qquad [4.63]$$

REMARK 4.3.– As results from [4.61] and [4.62], the knowledge of the kernel **Q** is necessary and sufficient to solve the evolution equations [4.63], Markov evolution for backward times. The introduction of backward recurrence times gives greater information on the studied system without the necessity of new statistical data.

4.8.2. *Semi-Markov backward processes in discrete time for the non-homogeneous case*

Now, we introduce the non-homogeneous backward SMP.

A time axis is reported to explain the backward introduction in Figure 4.6.

As far as the non-homogeneous SMP is concerned, we know that at time s, the system entered in the state i and then the probability of being in the state j at time t is given by $\phi_{ij}(s,t)$. In the non-homogeneous environment, the backward process considers that the system entered the state i at time $u=s-l$ and that it remained in the state i up to time s. The backward time, which highlights how much time before the system entered i, is $l = s - T_n$. The transition probabilities are conditioned by the entrance time into the state i and to the fact that there are no transitions in the system up to the time s. Under this hypothesis, relations [4.61] and [4.62] are rewritten in the following way:

$$D_{ij}(l,s;t)=\begin{cases}\dfrac{1-H_i(l,t)}{1-H_i(l,s)} & \text{if} \quad i=j,\\ 0 & \text{if} \quad i\neq j,\end{cases} \qquad [4.64]$$

$$b_{ij}(l,s;t)=\frac{b_{ij}(l,t)}{1-H_i(l,s)}, \qquad [4.65]$$

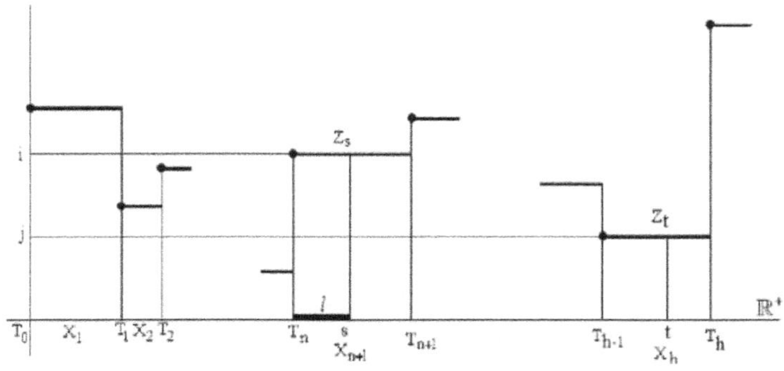

Figure 4.6. *Backward time axis*

As for HSMP, the backward process value at time t noted $B(t)$, is given by

$$B(t)=t-T_{N(t)}, \qquad [4.66]$$

we introduce as in the continuous case, the following transition probabilities

$${}^b\phi_{ij}(l,s;t)=P(Z(t)=j\mid Z(s)=i,B(s)=l) \qquad [4.67]$$

i.e. semi-Markov transition probabilities with a backward value at starting time s equal to l.

Bearing in mind the following equivalence between events

$$\{Z(s)=i,C(s)=l\}\equiv\{J_{N(s)}=i,T_{N(s)}=s-l,T_{N(s)+1}>s\}$$

it is possible to prove that the evolution equation of the discrete time non-homogeneous backward SMP is:

$$ {}^{b}\phi_{ij}(l,s;t) = D_{ij}(l,s;t) + \sum_{\beta\in I}\sum_{\vartheta=s+1}^{t} \phi_{\beta j}(\vartheta,t) b_{i\beta}(l,s;\vartheta) . \quad [4.68] $$

At the moment, we only report the definitions of mono-unireducible matrix, mono-unireducible SMP and then we derive some connections and results for mono-unireducible backward SMP.

DEFINITION 4.5.– *Let*

$$ \mathbf{A} = \begin{bmatrix} \mathbf{A}_{1,1} & \mathbf{A}_{1,2} \\ \mathbf{0}^{\mathrm{T}} & \mathbf{A}_{2,2} \end{bmatrix} $$

be a $m \times m$ non-negative matrix, where:

– $\mathbf{A}_{1,1}$ is a m–1×m–1 circuit matrix (see D'Amico *et al.* [D'AM 08]) (i.e. irreducible matrix with all the elements on the main diagonal greater than 0),

– $\mathbf{A}_{1,2}$ is a m–1 non-negative column vector in which at least 1 element is positive, – $\mathbf{A}_{2,2} = [a_{mm}]$, $a_{mm} > 0$ and $\mathbf{0}^{\mathrm{T}}$ is a m–1 null row vector;

then **A** is called a mono-unireducible matrix.

DEFINITION 4.6 (*D'Amico et al. 2006*).– *A discrete time non-homogeneous semi-Markov process is mono-unireducible if:*

1) $\forall s \in \mathbb{N} : \mathbf{B}(s;s+1) = \left(b_{ij}(s,s+1)\right)_{i,j\in I}$ is mono-unireducible,

2) $\forall i \in I, \forall s < t; s,t \in \mathbb{N} : b_{ii}(s;t) > 0$.

DEFINITION 4.7.– *A discrete time non-homogeneous backward semi-Markov process is mono-unireducible if:*

1) $\forall l,s \in \mathbb{N} : \mathbf{B}(l,s;s+1) = \left(b_{ij}(l,s;s+1)\right)_{i,j\in I}$ is mono-unireducible,

2) $\forall i \in I, \forall l,s < t;\ l,s,t \in \mathbb{N} : b_{ii}(l,s;t) > 0$.

LEMMA 4.1.– A discrete time non-homogeneous backward semi-Markov process $(Z(s), C(s))$ is mono-unireducible if, and only if, the corresponding discrete time non-homogeneous semi-Markov process $Z(s)$ is mono-unireducible.

PROPOSITION 4.2.– Let ${}^b\phi_{ij}(l,s;t)$ be the evolution equation of a discrete time non-homogeneous mono-unireducible backward SMP. Then it results:

$$\lim_{t\to\infty} {}^b\phi_{ij}(l;s,t) = \lim_{t\to\infty} \phi_{ij}(s,t) = \begin{cases} 1 & \text{if} \quad j = m \\ 0 & \text{if} \quad j \neq m \end{cases}$$

4.8.3. *DTSMP numerical solutions*

The evolution equations of the (DTHSMP) and (DTNHSMP) as defined in relations [4.73a] and [4.73b] are the following:

$$\phi_{ij}(k) = d_{ij}(k) + \sum_{l=1}^{m}\sum_{\tau=1}^{k} \phi_{lj}(k-\tau) b_{il}(\tau), \tag{4.69}$$

$$\phi_{ij}(u,k) = d_{ij}(u,k) + \sum_{l=1}^{m}\sum_{\tau=u+1}^{k} \phi_{lj}(\tau,k) b_{il}(u,\tau). \tag{4.70}$$

These equations can be rewritten in matrix form as follows:

$$\mathbf{\Phi}(k) = \mathbf{D}(k) + \sum_{\tau=1}^{k} \mathbf{B}(\tau) * \mathbf{\Phi}(k-\tau),$$

$$\mathbf{\Phi}(u,k) = \mathbf{D}(u,k) + \sum_{\tau=u+1}^{k} \mathbf{B}(u,\tau) * \mathbf{\Phi}(\tau,k),$$

or equivalently:

$$\mathbf{\Phi}(k) - \sum_{\tau=1}^{k} \mathbf{B}(\tau) * \mathbf{\Phi}(k-\tau) = \mathbf{D}(k), \tag{4.71}$$

$$\mathbf{\Phi}(u,k) - \sum_{\tau=u+1}^{k} \mathbf{B}(\tau,k) * \mathbf{\Phi}(\tau,k) = \mathbf{D}(u,k), \ k \in \mathbb{N}, u \leq k. \tag{4.72}$$

Taking into account that $k \in \mathbb{N}$, both equations [4.71] and [4.72] can be written more compactly as:

$$\mathbf{U} * \mathbf{\Phi} = \mathbf{D}. \qquad [4.73]$$

For the homogeneous case, it results that:

$$\mathbf{U} = \begin{bmatrix} \mathbf{I} & \mathbf{0} & \mathbf{0} & \mathbf{0} & \cdots \\ -\mathbf{B}(1) & \mathbf{I} & \mathbf{0} & \mathbf{0} & \cdots \\ -\mathbf{B}(2) & -\mathbf{B}(1) & \mathbf{I} & \mathbf{0} & \cdots \\ -\mathbf{B}(3) & -\mathbf{B}(2) & -\mathbf{B}(1) & \mathbf{I} & \cdots \\ \vdots & \vdots & \vdots & \vdots & \ddots \end{bmatrix}, \qquad [4.74]$$

$$\mathbf{\Phi} = \begin{bmatrix} \mathbf{\Phi}(0) \\ \mathbf{\Phi}(1) \\ \mathbf{\Phi}(2) \\ \mathbf{\Phi}(3) \\ \vdots \end{bmatrix}, \mathbf{D} = \begin{bmatrix} \mathbf{D}(0) \\ \mathbf{D}(1) \\ \mathbf{D}(2) \\ \mathbf{D}(3) \\ \vdots \end{bmatrix}$$

and in the non-homogeneous case:

$$\mathbf{U} = \begin{bmatrix} \mathbf{I} & -\mathbf{B}(0,1) & -\mathbf{B}(0,2) & -\mathbf{B}(0,3) & \cdots \\ \mathbf{0} & \mathbf{I} & -\mathbf{B}(1,2) & -\mathbf{B}(1,3) & \cdots \\ \mathbf{0} & \mathbf{0} & \mathbf{I} & -\mathbf{B}(2,3) & \cdots \\ \mathbf{0} & \mathbf{0} & \mathbf{0} & \mathbf{I} & \cdots \\ \vdots & \vdots & \vdots & \vdots & \ddots \end{bmatrix}, \qquad [4.75]$$

$$\mathbf{D} = \begin{bmatrix} \mathbf{I} & \mathbf{D}(0,1) & \mathbf{D}(0,2) & \mathbf{D}(0,3) & \cdots \\ \mathbf{0} & \mathbf{I} & \mathbf{D}(1,2) & \mathbf{D}(1,3) & \cdots \\ \mathbf{0} & \mathbf{0} & \mathbf{I} & \mathbf{D}(2,3) & \cdots \\ \mathbf{0} & \mathbf{0} & \mathbf{0} & \mathbf{I} & \cdots \\ \vdots & \vdots & \vdots & \vdots & \ddots \end{bmatrix}.$$

The following theorem holds for both homogeneous and non-homogeneous cases.

THEOREM 4.1 – Equations [4.71] and [4.72] admit a unique solution.

PROOF.– The determinant of matrix $\mathbf{U}$ is absolutely convergent (Riesz [RIE 13]); more precisely $\det(\mathbf{U}) = 1$ and consequently matrix $\mathbf{U}$ is invertible. □

In the homogeneous case, [4.71] can be seen as an infinite linear system with an infinite number of unknowns. It ensues from theorem 4.1 that the system is solvable.

To solve such a system, it is usually necessary to apply the truncation method Riesz [RIE 13]; but in our case, it is very simple to find the solution.

Obviously, $\mathbf{\Phi}(0) = \mathbf{I}$ and, once $\mathbf{\Phi}(0)$ is known we obtain:

$$\mathbf{\Phi}(1) = \mathbf{B}(1) * \mathbf{\Phi}(0) + \mathbf{D}(1) .$$

Once $\mathbf{\Phi}(0), \mathbf{\Phi}(1), \ldots, \mathbf{\Phi}(k)$ are known, then:

$$\mathbf{\Phi}((k+1)) = \sum_{\tau=1}^{k+1} \mathbf{B}(\tau) * \mathbf{\Phi}(k+1-\tau) + \mathbf{D}(k+1) \qquad [4.76]$$

and it is not necessary to apply the truncation method. Furthermore, these results are obtained without any matrix inversion.

As in the homogeneous case, equation [4.71] can be seen as an infinite linear system with an infinite number of unknowns, but the non-homogeneous matrix equation is different from the homogeneous equation because the coefficient matrix is upper triangular in the non-homogeneous case and lower triangular in the homogeneous.

Also in this case, the truncation method should not be applied. In the homogeneous case, this result is trivial. In the non-homogeneous case, the result is not so immediate. We report, with more precision, the following result given in Janssen & Manca [JAN 01].

PROPOSITION 4.3.– The solution of the infinite order linear system [4.72] can be given explicitly step by step.

PROOF.– For the proof see Janssen & Manca [JAN 01]. □

REMARK 4.4.– Here it should also be mentioned that the upper triangularity of the coefficient block matrix and the fact that the matrices on the main diagonal are identity matrices implies that the results are obtained with no matrix inversion.

The following two theorems hold for both the homogeneous and non-homogeneous cases but only the homogeneous results will be given.

A straightforward proof of the randomness of the matrix $\mathbf{\Phi}$ is possible by exploiting the following theorem.

THEOREM 4.2.– It was put in order $\mathbf{\Phi}(k)$ are stochastic.

PROOF.– The result is true for $\mathbf{\Phi}(0)=\mathbf{I}$. We suppose that it is true for $\tau=1,\ldots,k$. We have to check what happens under these hypotheses for $k+1$.

From [4.69], equation [4.76] becomes:

$$\mathbf{\Phi}(k+1)=\sum_{\tau=1}^{k+1}\left(\mathbf{Q}(\tau)-\mathbf{Q}(\tau-1)\right)\mathbf{\Phi}(k+1-\tau)+\mathbf{D}(k+1). \qquad [4.77]$$

To prove that $\mathbf{\Phi}(k+1)$ is stochastic, we have to show that:

$$\sum_{j=1}^{m}\phi_{ij}(k+1)=1, \quad i=1,\ldots,\ m.$$

An element of [4.77] is given by:

$$\phi_{ij}(k+1)=d_{ij}(k+1)+\sum_{\tau=1}^{k+1}\sum_{l=1}^{m}Q_{il}(\tau)\phi_{lj}(k+1-\tau)$$
$$-\sum_{\tau=1}^{k+1}\sum_{l=1}^{m}Q_{il}(\tau-1)\ \ \phi_{lj}(k+1-\tau).$$

$$\sum_{j=1}^{m}\phi_{ij}(k+1)=1-H_i(k+1)+\sum_{j=1}^{m}\sum_{\tau=1}^{k+1}\sum_{l=1}^{m}Q_{il}(\tau)\phi_{lj}(k+1-\tau)$$
$$-\sum_{j=1}^{m}\sum_{\tau=1}^{k+1}\sum_{l=1}^{m}Q_{il}(\tau-1)\ \ \phi_{lj}(k+1-\tau),$$

$$\sum_{j=1}^{m}\phi_{ij}(k+1)=1-\sum_{j=1}^{m}Q_{ij}(k+1)+\sum_{\tau=1}^{k+1}\sum_{l=1}^{m}Q_{il}(\tau)\sum_{j=1}^{m}\phi_{lj}(k+1-\tau)$$
$$-\sum_{\tau=1}^{k+1}\sum_{l=1}^{m}Q_{il}(\tau-1)\sum_{j=1}^{m}\phi_{lj}(k+1-\tau),$$

For the inductive hypothesis, it results:

$$\sum_{j=1}^{m}\phi_{ij}(k+1)=1-\sum_{j=1}^{m}Q_{ij}(k+1)+\sum_{\tau=1}^{k+1}\sum_{l=1}^{m}Q_{il}(\tau)$$
$$-\sum_{\tau=1}^{k+1}\sum_{l=1}^{m}Q_{il}(\tau-1),$$

That means: $\sum_{j=1}^{m}\phi_{ij}(k+1)=1$ □

4.9. Discrete time reward processes

In this section, as in Chapter 3 in the Markov case, we will present undiscounted and discounted discrete time semi-Markov reward processes.

In the semi-Markov non-homogeneous case, it is possible to write more than 200 different evolution equations of semi-Markov reward processes (SMRWP). We develop only three cases, the simplest and the most general. For a wider approach, the reader can refer to Janssen-Manca [JAN 06, JAN 07].

4.9.1. *Undiscounted SMRWP*

For each given case, we will present the immediate and the due cases, both in homogeneous and non-homogeneous environments. We will give first the simplest case (only with permanence rewards and fixed rate of interest and rewards) and after the general ones. The same cases will be given for discounted processes.

4.9.1.1. *Fixed permanence rewards*

We assume that:

1) rewards are fixed in time,

2) rewards are given only for permanence in the state.

First we present the immediate case.

$V_i(t)$ $(V_i(s,t))$ represents the *Mean Total Reward* (MTR) paid or received in t periods (from time s to time t), given that at time 0 (at time s) the system was in state i.

At time 1, the evolution equation for the homogeneous immediate case is given by the following relation:

$$V_i(1) = \left(1 - H_i(1)\right)\psi_i + \sum_{k=1}^{m} b_{ik}(1)\psi_i + \sum_{k=1}^{m}\sum_{\vartheta=0}^{1} b_{ik}(\vartheta)V_k(1-\vartheta). \qquad [4.78]$$

To have a good understanding of the evolution equation, let us first say that relation [4.53] implies that $b_{ij}(1) = Q_{ij}(1)$ and so relation [4.78] can be decomposed in the following way:

$$V_i(1) = \left(1 - \sum_{k=1}^{m} Q_{ik}(1)\right)\psi_i + \sum_{k=1}^{m} Q_{ik}(1)\psi_i + \sum_{k=1}^{m}\sum_{\vartheta=0}^{1} Q_{ik}(\vartheta)V_k(1-\vartheta)$$

where $V_k(0) = 0,\ \forall k$ and $Q_{ik}(0) = 0\ \forall i,k$, and so:

$$V_i(1) = \psi_i.$$

For the next step, we can write that:

$$V_i(2) = (1 - H_i(t))\psi_i 2 + \sum_{k=1}^{m}\sum_{\vartheta=1}^{2} b_{ik}(\vartheta)\psi_i\vartheta + \sum_{k=1}^{m}\sum_{\vartheta=1}^{2} b_{ik}(\vartheta)V_k(2-\vartheta). \qquad [4.79]$$

This time, two rewards must be paid but in different ways. We divide the evolution equation into the following three parts:

– the term $(1 - H_i(t))\psi_i$ represents the rewards obtained without state changes;

– the expression $\sum_{k=1}^{m}\sum_{\vartheta=1}^{2} b_{ik}(\vartheta)\psi_i\vartheta$ gives the rewards obtained before the change of state. As $b_{ik}(0) = 0\ \forall i,k$, the sum on ϑ begins from 1;

– the double sum $\sum_{k=1}^{m}\sum_{\vartheta=1}^{2} b_{ik}(\vartheta)V_k(2-\vartheta)$ gives the rewards paid or earned after the transitions.

For time t, we get the following general result:

$$V_i(t) = (1-H_i(t))\psi_i t + \sum_{k=1}^{m}\sum_{\vartheta=1}^{t} b_{ik}(\vartheta)\psi_i \vartheta + \sum_{k=1}^{m}\sum_{\vartheta=1}^{t} b_{ik}(\vartheta)V_k(t-\vartheta).$$

The general formula in the non-homogeneous case is:

$$V_i(s,t) = (1-H_i(s,t))(t-s)\psi_i + \sum_{k=1}^{m}\sum_{\vartheta=s+1}^{t} b_{ik}(s,\vartheta)(\vartheta-s)\psi_i + \sum_{k=1}^{m}\sum_{\vartheta=s+1}^{t} b_{ik}(s,\vartheta)V_k(\vartheta,t). \quad [4.80]$$

In this simple case, the due and the immediate processes correspond.

4.9.1.2. *Variable permanence and transition rewards*

Here we assume that:

1) rewards are variable in time,

2) rewards are given for permanence in the state and at a given transition.

Under these hypotheses, we get respectively for homogeneous and non-homogeneous environments, in the *immediate cases*, the following results:

$$V_i(t) = (1-H_i(t))\sum_{\tau=1}^{t}\psi_i(\tau) + \sum_{k=1}^{m}\sum_{\vartheta=1}^{t} b_{ik}(\vartheta)\sum_{\tau=1}^{\vartheta}\psi_i(\tau) + \sum_{k=1}^{m}\sum_{\vartheta=1}^{t} b_{ik}(\vartheta)\gamma_{ik}(\vartheta) + \sum_{k=1}^{m}\sum_{\vartheta=1}^{t} b_{ik}(\vartheta)V_k(t-\vartheta), \quad [4.81]$$

$$V_i(s,t) = (1-H_i(s,t))\sum_{\tau=s+1}^{t}\psi_i(\tau) + \sum_{k=1}^{m}\sum_{\vartheta=s+1}^{t} b_{ik}(s,\vartheta)\sum_{\tau=1}^{\vartheta}\psi_i(\tau) + \sum_{k=1}^{m}\sum_{\vartheta=s+1}^{t} b_{ik}(s,\vartheta)\gamma_{ik}(\vartheta) + \sum_{k=1}^{m}\sum_{\vartheta=s+1}^{t} b_{ik}(s,\vartheta)V_k(\vartheta,t). \quad [4.82]$$

In the *due case* we obtain:

$$\begin{aligned}\ddot{V}_i(t) &= (1-H_i(t))\sum_{\tau=1}^{t}\psi_i(\tau-1)+\sum_{k=1}^{m}\sum_{\vartheta=1}^{t}b_{ik}(\vartheta)\sum_{\tau=1}^{\vartheta}\psi_i(\tau-1)\\&+\sum_{k=1}^{m}\sum_{\vartheta=1}^{t}b_{ik}(\vartheta)\gamma_{ik}(\vartheta)+\sum_{k=1}^{m}\sum_{\vartheta=1}^{t}b_{ik}(\vartheta)\ddot{V}_k(t-\vartheta),\end{aligned} \tag{4.83}$$

$$\begin{aligned}\ddot{V}_i(s,t) &= (1-H_i(s,t))\sum_{\tau=s+1}^{t}\psi_i(\tau-1)+\sum_{k=1}^{m}\sum_{\vartheta=s+1}^{t}b_{ik}(s,\vartheta)\gamma_{ik}(\vartheta)\\&+\sum_{k=1}^{m}\sum_{\vartheta=s+1}^{t}b_{ik}(s,\vartheta)\sum_{\tau=1}^{\vartheta}\psi_i(\tau-1)+\sum_{k=1}^{m}\sum_{\vartheta=s+1}^{t}b_{ik}(s,\vartheta)\ddot{V}_k(\vartheta,t).\end{aligned} \tag{4.84}$$

The difference between immediate and due is given only by the time of payment of the rewards.

The presence of the lump sums given or taken at the moment of transition times is taken into consideration.

4.9.1.3. *Non-homogeneous permanence and transition rewards*

In the last immediate case model, the rewards are non-homogeneous and so we have to consider only the non-homogeneous case.

Assumptions are thus:

1) rewards depend on the times s and t,

2) permanence and transition rewards are non-homogeneous.

Here, only the non-homogeneous case makes sense and the evolution equations take the form:

$$\begin{aligned}V_i(s,t) &= (1-H_i(s,t))\sum_{\tau=s+1}^{t}\psi_i(s,\tau)+\sum_{k=1}^{m}\sum_{\vartheta=s+1}^{t}b_{ik}(s,\vartheta)\sum_{\tau=s+1}^{\vartheta}\psi_i(s,\tau)\\&+\sum_{k=1}^{m}\sum_{\vartheta=s+1}^{t}b_{ik}(s,\vartheta)\gamma_{ik}(s,\vartheta)+\sum_{k=1}^{m}\sum_{\vartheta=s+1}^{t}b_{ik}(s,\vartheta)V_k(\vartheta,t).\end{aligned} \tag{4.85}$$

$$\begin{aligned}\ddot{V}_i(s,t) &= (1-H_i(s,t))\sum_{\tau=s+1}^{t}\psi_i(s,\tau-1)+\sum_{k=1}^{m}\sum_{\vartheta=s+1}^{t}b_{ik}(s,\vartheta)\sum_{\tau=1}^{\vartheta}\psi_i(s,\tau-1)\\&+\sum_{k=1}^{m}\sum_{\vartheta=s+1}^{t}b_{ik}(s,\vartheta)\gamma_{ik}(s,\vartheta)+\sum_{k=1}^{m}\sum_{\vartheta=s+1}^{t}b_{ik}(s,\vartheta)\ddot{V}_k(\vartheta,t).\end{aligned} \tag{4.86}$$

The other non-discounted cases can be treated in a similar way and are left to the reader, who can refer also to Janssen& Manca [JAN 06, JAN 07].

4.9.2. *Discounted SMRWP*

For the discounted case developed in this section, we assume that all the rewards are discounted at time 0 in the homogeneous case and at time s in the non-homogeneous case. Let us point out that these models are very important for insurance applications.

4.9.2.1. *Fixed permanence and interest rate cases*

In the first formulation of this case we suppose that:

1) rewards are fixed in time,

2) rewards are given only for permanence in the state,

3) interest rate r is fixed.

In this case, $V_i(t)$ represents the *Rewards Mean Present Value* (RMPV) paid or received in a time t, given that at time 0 the system is in state i.

Under these hypotheses, a similar reasoning as before leads to the following result for the evolution equation, first for the homogeneous immediate case:

$$V_i(1) = \left(1-H_i(1)\right)\psi_i\nu^1+\sum_{k=1}^{m}b_{ik}(1)\psi_i\nu^1+\sum_{k=1}^{m}\sum_{\vartheta=1}^{1}b_{ik}(\vartheta)V_k(1-\vartheta)\nu^1=\psi_i\nu^1,$$

$$V_i(t) = (1-H_i(t))\psi_i a_{\overline{t}|r}+\sum_{k=1}^{m}\sum_{\vartheta=1}^{t}b_{ik}(\vartheta)\psi_i a_{\overline{\vartheta}|r}+\sum_{k=1}^{m}\sum_{\vartheta=1}^{t}b_{ik}(\vartheta)V_k(t-\vartheta)\nu^{\vartheta}. \qquad [4.87]$$

For the non-homogeneous case, this last result becomes:

$$\begin{aligned}V_i(s,t) &= (1-H_i(s,t))\psi_i a_{\overline{t-s}|r}+\sum_{k=1}^{m}\sum_{\vartheta=s+1}^{t}b_{ik}(s,\vartheta)\psi_i a_{\overline{\vartheta-s}|r}\\&+\sum_{k=1}^{m}\sum_{\vartheta=s+1}^{t}b_{ik}(s,\vartheta)V_k(\vartheta,t)\nu^{\vartheta-s}.\end{aligned} \qquad [4.88]$$

To explain these results, we divide the evolution equation into three parts. The meaning is the same given in the previous cases but we use annuity formulas.

Let us just make the following comments:

The term $(1-H_i(s,t))\psi_i a_{\overline{t-s}|}$ represents the present value of the rewards obtained without state changes. More precisely $(1-H_i(s,t))$ is the probability of remaining in the state i and $\psi_i a_{\overline{t-s}|}$ is the present value of a constant annuity of $t-s$ payments ψ_i.

The term $\sum_{k=1}^{m}\sum_{\vartheta=s+1}^{t} b_{ik}(s,\vartheta)\psi_i a_{\overline{\vartheta-s}|}$ gives the present value of the rewards obtained before the change of state.

The term $\sum_{k=1}^{m}\sum_{\vartheta=s+1}^{t} b_{ik}(s,\vartheta)V_k(\vartheta,t)\nu^{\vartheta-s}$ gives the present value of the rewards paid or earned after the transitions and as the change of state happens at time ϑ, it is necessary to discount the reward values at time s.

In the *due environment* we obtain:

$$\ddot{V}_i(t) = (1-H_i(t))\psi_i\ddot{a}_{\overline{t}|} + \sum_{k=1}^{m}\sum_{\vartheta=1}^{t} b_{ik}(\vartheta)\psi_i\ddot{a}_{\overline{\vartheta}|} + \sum_{k=1}^{m}\sum_{\vartheta=1}^{t} b_{ik}(\vartheta)\ddot{V}_k(t-\vartheta)\nu^{\vartheta-1} \qquad [4.89]$$

$$\begin{aligned}\ddot{V}_i(s,t) &= (1-H_i(s,t))\psi_i\ddot{a}_{\overline{t-s}|} + \sum_{k=1}^{m}\sum_{\vartheta=s+1}^{t} b_{ik}(s,\vartheta)\psi_i\ddot{a}_{\overline{\vartheta-s}|} \\ &\quad + \sum_{k=1}^{m}\sum_{\vartheta=s+1}^{t} b_{ik}(s,\vartheta)\ddot{V}_k(\vartheta,t)\nu^{\vartheta-s-1}.\end{aligned} \qquad [4.90]$$

4.9.2.2. *Variable interest rate, permanence and transition cases*

Now we make the following assumptions:

1) rewards are variable in time,

2) rewards are given for permanence in the state and at a given transition,

3) the interest rate is variable.

Under these hypotheses, in the immediate case we get the following relations:

$$\begin{aligned}V_i(t) &= (1-H_i(t))\sum_{h=1}^{t}\psi_i(h)\nu(h) + \sum_{k=1}^{m}\sum_{\vartheta=1}^{t} b_{ik}(\vartheta)\sum_{h=1}^{\vartheta}\psi_i(h)\nu(h) \\ &\quad + \sum_{k=1}^{m}\sum_{\vartheta=1}^{t} b_{ik}(\vartheta)\gamma_{ik}(\vartheta)\nu(\vartheta) + \sum_{k=1}^{m}\sum_{\vartheta=1}^{t} b_{ik}(\vartheta)V_k(t-\vartheta)\nu(\vartheta),\end{aligned} \qquad [4.91]$$

$$
\begin{aligned}
V_i(s,t) &= (1-H_i(s,t))\sum_{h=s+1}^{t}\psi_i(h)\nu(s,h)+\sum_{k=1}^{m}\sum_{\vartheta=s+1}^{t}b_{ik}(s,\vartheta)\sum_{h=s+1}^{\vartheta}\psi_i(h)\nu(s,h)\\
&+\sum_{k=1}^{m}\sum_{\vartheta=s+1}^{t}b_{ik}(s,\vartheta)\gamma_{ik}(\vartheta)\nu(s,\vartheta)+\sum_{k=1}^{m}\sum_{\vartheta=s+1}^{t}b_{ik}(s,\vartheta)V_k(\vartheta,t)\nu(s,\vartheta).
\end{aligned}
\quad [4.92]
$$

In the *due case* we get:

$$
\begin{aligned}
\ddot{V}_i(t) &= (1-H_i(t))\sum_{\tau=0}^{t-1}\psi_i(\tau)\nu(\tau)+\sum_{k=1}^{m}\sum_{\vartheta=1}^{t}b_{ik}(\vartheta)\sum_{\tau=0}^{\vartheta-1}\psi_i(\tau)\nu(\tau)\\
&+\sum_{k=1}^{m}\sum_{\vartheta=1}^{t}\nu(\vartheta-1)b_{ik}(\vartheta)\ddot{V}_k(t-\vartheta)+\sum_{k=1}^{m}\sum_{\vartheta=1}^{t}\nu(\vartheta)b_{ik}(\vartheta)\gamma_{ik}(\vartheta),
\end{aligned}
\quad [4.93]
$$

$$
\begin{aligned}
\ddot{V}_i(s,t) &= (1-H_i(s,t))\sum_{\tau=s}^{t-1}\psi_i(\tau)\nu(s,\tau)+\sum_{k=1}^{m}\sum_{\vartheta=s+1}^{t}b_{ik}(s,\vartheta)\sum_{\tau=s}^{\vartheta-1}\psi_i(\tau)\nu(s,\tau)\\
&+\sum_{k=1}^{m}\sum_{\vartheta=s+1}^{t}\nu(s,\vartheta-1)b_{ik}(s,\vartheta)\ddot{V}_k(\vartheta,t)+\sum_{k=1}^{m}\sum_{\vartheta=s+1}^{t}\nu(s,\vartheta)b_{ik}(s,\vartheta)\gamma_{ik}(\vartheta).
\end{aligned}
\quad [4.94]
$$

4.9.2.3. *Non-homogeneous interest rate, permanence and transition case*

For our last case, we consider non-homogeneous rewards and interest rate. And so the basic assumptions are:

1) rewards are non-homogeneous,

2) rewards are given for premanence and transitions,

3) interest rate is non-homogeneous.

It can easily be verified that the evolution equations take the form:

$$
\begin{aligned}
V_i(s,t) &= (1-H_i(s,t))\sum_{\tau=s+1}^{t}\psi_i(s,\tau)\dot{\nu}(s,\tau)+\sum_{k=1}^{m}\sum_{\vartheta=s+1}^{t}b_{ik}(s,\vartheta)V_k(\vartheta,t)\dot{\nu}(s,\vartheta)\\
&+\sum_{k=1}^{m}\sum_{\vartheta=s+1}^{t}b_{ik}(s,\vartheta)\gamma_{ik}(s,\vartheta)\dot{\nu}(s,\vartheta)+\sum_{k=1}^{m}\sum_{\vartheta=s+1}^{t}b_{ik}(s,\vartheta)\sum_{\tau=s+1}^{\vartheta}\psi_i(s,\tau)\dot{\nu}(s,\tau),
\end{aligned}
\quad [4.95]
$$

$$
\begin{aligned}
\ddot{V}_i(s,t) &= (1-H_i(s,t))\sum_{\tau=s}^{t-1}\psi_i(s,\tau)\dot{\nu}(s,\tau)+\sum_{k=1}^{m}\sum_{\vartheta=s+1}^{t}\dot{\nu}(s,\vartheta)b_{ik}(s,\vartheta)\gamma_{ik}(s,\vartheta)\\
&+\sum_{k=1}^{m}\sum_{\vartheta=s+1}^{t}\dot{\nu}(s,\vartheta-1)b_{ik}(s,\vartheta)\ddot{V}_k(\vartheta,t)+\sum_{k=1}^{m}\sum_{\vartheta=s+1}^{t}b_{ik}(s,\vartheta)\sum_{\tau=0}^{\vartheta-1}\psi_i(s,\tau)\dot{\nu}(s,\tau),
\end{aligned}
\quad [4.96]
$$

4.9.3. *General algorithms for DTSMRWP*

In the previous section, we presented useful discrete time semi-Markov reward processes as well as general global models for which the evolution equations can be written in the matrix form

$$\mathbf{U}*\mathbf{V}=\mathbf{C}. \qquad [4.97]$$

In the homogeneous case, **U** is an infinite order lower-triangular matrix whose elements are $m \times m$ matrices and **V** and **C** are infinite order vectors whose elements are m-dimensional vectors.

In the non-homogeneous case in [4.95], **U** is an infinite order upper-triangular matrix whose elements are $m \times m$ matrices and **V** and **C** are infinite order matrices whose elements are m-dimensional vectors.

Of course, matrices **U** and **C** depend on the particular models presented in the preceding section.

For real-life applications, it is generally sufficient to study the problem on a finite time horizon $[0,T]$ and then the infinite system [4.95] becomes a finite system

$${}^{T}\mathbf{U}*{}^{T}\mathbf{V}={}^{T}\mathbf{C}$$

where ${}^{T}\mathbf{U}$ is a square lower triangular block matrix of order T+1 in the homogeneous case and an upper triangular block matrix in the non-homogeneous case. ${}^{T}\mathbf{C}$, ${}^{T}\mathbf{V}$ are respectively T+1-dimensional vectors, in the homogeneous case, and matrices, in the non-homogeneous case, whose elements are m-dimensional vectors.

We will present briefly two general algorithms (homogeneous and non-homogeneous) solving all possible reward cases.

The main steps of these algorithms are the following:

(1) Homogeneous case

Input – selectors that choose among the SMRWP, the number of states and the number of periods, the permanence and transition rewards, the fixed or variable interest rate, the transition matrix **P** and the matrix ${}^{T}\mathbf{F}$ of waiting time d.f.

CONSTRUCT – ${}^{T}\mathbf{Q}$

CONSTRUCT – ${}^{T}\mathbf{B}$

Construct – ${}^{T}\mathbf{H}$

Construct – ${}^{T}\mathbf{D}$

Construct – the permanence rewards

Construct – the transition rewards

Construct – the vector discount factors

Construct – ${}^{T}\mathbf{C}$, known terms

Solve – the system and find ${}^{T}\mathbf{V}$

(2) Non-homogeneous case

Input – selectors, the number of states and the number of periods, the permanence and transition rewards, the fixed, variable or non-homogeneous interest rate, the transition matrix **P** and the matrix ${}^{T}\mathbf{F}$ waiting time d.f.

CONSTRUCT – ${}^{T}\mathbf{Q}$

CONSTRUCT – ${}^{T}\mathbf{B}$

Construct – ${}^{T}\mathbf{H}$

Construct – ${}^{T}\mathbf{D}$

Construct – the permanence rewards

Construct – the transition rewards

Construct – the matrix discount factors

Construct – ${}^{T}\mathbf{C}$, known terms

Solve – the system and find ${}^{T}\mathbf{V}$

These algorithms are able to solve any DTSMRWP. They constitute a very important tool for the application of semi-Markov reward processes in many applied sciences.

4.10. Markov renewal functions in the homogeneous case

4.10.1. *Entrance times*

This section will extend result of the section 3.3 concerning renewal functions to MRP.

Let us consider a homogeneous MRP of kernel **Q** and to avoid trivialities, we will assume that:

$$\sup_{i,j} Q_{ij}(0) < 1, \quad [4.98]$$

where the functions Q_{ij} are defined by relation [4.7].

If the initial state J_0 is i, let us define the r.v. $T_n(i|i)$, $n \geq 1$, as the times (possibly infinite) of *successive returns* to state i, also called *successive entrance times* into $\{i\}$.

From the regenerative property of MRP, whenever the process enters into state i, say at time t, the evolution of the process on $[t, \infty)$ is probabilistically the same as if we had started at time 0 in the same state i.

It follows that the process $(T_n(i|i), n \geq 0)$ with:

$$T_0(i|i) = 0 \quad [4.99]$$

is a renewal process that could possibly be defective.

From now on, the r.v. $T_n(i|i)$ will be called the *nth return time to state i.*

More generally, let us also fix state j, different from the state i already fixed; we can also define the *nth return* or *entrance time to state j*, but *starting from i as the initial state*. This time, possibly infinite as well, will be represented by $(T_n(j|i), n \geq 0)$, also using the convention that

$$T_0(j|i) = 0. \quad [4.100]$$

Now, the sequence $\left(T_n\left(j|i\right), n \geq 0\right)$ is a delayed renewal process with values in $\mathbb{R}^+$.

It is thus defined by two d.fs.: G_{ij} being that of $T_1\left(j|i\right)$ and G_{jj} that of $T_2\left(j|i\right) - T_1\left(j|i\right)$, so that:

$$\begin{aligned} G_{ij}(t) &= P\left(T_1\left(j|i\right) \leq t\right), \\ G_{jj}(t) &= P\left(T_n\left(j|i\right) - T_{n-1}\left(j|i\right) \leq t\right), \; n \geq 2. \end{aligned} \qquad [4.101]$$

Of course, the d.f.G_{jj} suffices to define the renewal process $\left(T_n\left(j|j\right), \; n \geq 0\right)$.

REMARK 4.5.– From the preceding definitions, we can also write that:

$$\begin{aligned} & G_{ij}(t) = P\left(N_j(t) > 0 | J_0 = i\right); \; i, j \in I, \\ & P\left(T_1\left(j|i\right) = +\infty\right) = 1 - G_{ij}(+\infty) \end{aligned} \qquad [4.102]$$

and for the mean of the $T_n\left(i|i\right)$, $n \geq 1$, possibly infinite, we obtain:

$$\mu_{ij} = E\left(T_1\left(j|i\right)\right) = \int_0^{\infty} t dG_{ij}(t), \qquad [4.103]$$

with the usual convention that

$$0 * +\infty = 0 \qquad [4.104]$$

The means $\mu_{ij}, i, j \in I$ are called the *first entrance* or *average return times*.

It is possible to show that the functions $G_{ij}, i, j \in I$ satisfy the following relationships:

$$G_{ij}(t) = \sum_{k=1}^{m} G_{kj} \bullet Q_{ik}(t) + (1 - G_{jj}) \bullet Q_{ij}(t), i, j \in I, t \geq 0. \qquad [4.105]$$

where the symbol $\bullet$ represents the classical Lebesgue-Stieltjes convolution product.

We will now define by A_{ij} and R_{ij} the associated *renewal functions*

$$A_{ij}(t)=E(N_j(t)|J_0=i),\\ R_{ij}(t)=E(N_j^{'}(t)|J_0=i) \quad [4.106]$$

where $N_j^{'}(t)=\delta_{ij}+N_j(t), j\in I$

so that

$$R_{ij}(t)=\delta_{ij}U_0(t)+A_{ij}(t). \quad [4.107]$$

Using the classical results of the renewal theory of Chapter 3, we obtain:

$$R_{jj}(t)=\sum_{n=0}^{+\infty}G_{jj}^{(n)}(t), j\in I,\\ R_{ij}(t)=G_{ij}\cdot R_{jj}(t) \quad [4.108]$$

or equivalently, we have:

$$R_{ij}(t)=\delta_{ij}U_0(t)+G_{ij}\bullet\sum_{n=0}^{\infty}G_{jj}^{(n)}(t),i,j\in I. \quad [4.109]$$

The following relations will express the renewal functions R_{ij}, $i,j\in I$ as a function of the kernel $\mathbf{Q}$ instead of the m^2 functions G_{ij}.

PROPOSITION 4.4.– For every i and j of I, we have:

$$R_{ij}(t)=\sum_{n=0}^{\infty}Q_{ij}^{(n)}(t). \quad [4.110]$$

Using matrix notation with:

$$\mathbf{R}=\left[R_{ij}\right], \quad [4.111]$$

relation [4.108] takes the form:

$$\mathbf{R}=\sum_{n=0}^{\infty}\mathbf{Q}^{(n)}. \quad [4.112]$$

Let us now introduce the L-S transform of matrices.

For any matrix of suitable functions A_{ij} from $\mathbb{R}^+$ to $\mathbb{R}$ represented by

$$\mathbf{A}=\left[A_{ij}\right] \qquad [4.113]$$

we will represent its L-S transform by:

$$\bar{\mathbf{A}}=\left[\bar{A}_{ij}\right] \qquad [4.114]$$

with

$$\bar{A}_{ij}(s)=\int_0^\infty e^{-st}dA_{ij}(t). \qquad [4.115]$$

Doing so for matrix $\mathbf{R}$, we obtain the matrix form of relation [4.110],

$$\bar{\mathbf{R}}(s)=\sum_{n=0}^{\infty}\left(\bar{\mathbf{Q}}(s)\right)^n. \qquad [4.116]$$

From this last relation, a simple algebraic argument shows that, for any $s>0$, relations

$$\bar{\mathbf{R}}(s)(\mathbf{I}-\bar{\mathbf{Q}}(s))=(\mathbf{I}-\bar{\mathbf{Q}}(s)\bar{\mathbf{R}}(s)=\mathbf{I} \qquad [4.117]$$

hold and so we also have:

$$\bar{\mathbf{R}}(s)=(\mathbf{I}-\bar{\mathbf{Q}}(s))^{-1}. \qquad [4.118]$$

We have thus proved the following proposition.

PROPOSITION 4.5.– The Markov renewal matrix $\mathbf{R}$ is given by

$$\mathbf{R}=\sum_{n=0}^{\infty}\mathbf{Q}^{(n)}, \qquad [4.119]$$

the series being convergent in $\mathbb{R}^+$, and the inverse existing for all positive s.

The knowledge of the Markov renewal matrix $\mathbf{R}$ or its L-S transform $\overline{\mathbf{R}}$ leads to useful expressions for d.f. of the first entrance times.

4.10.2. *The Markov renewal equation*

This section will extend the basic results related to the renewal equation developed in section 2.4 to the Markov renewal case.

Let us consider an MRP of kernel $\mathbf{Q}$.

From relation [4.110] we obtain:

$$\begin{aligned} R_{ij}(t) &= \delta_{ij}U_0(t) + \sum_{n=1}^{\infty} Q_{ij}^{(n)}(t) \\ &= \delta_{ij}U_0(t) + (Q \bullet R)_{ij}(t). \end{aligned} \quad [4.120]$$

Using matrix notation with:

$$\mathbf{I}(t) = \left[\delta_{ij}U_0(t)\right], \quad [4.121]$$

relations [4.120] take the form:

$$\mathbf{R}(t) = \mathbf{I}(t) + \mathbf{Q} \bullet \mathbf{R}(t). \quad [4.122]$$

This integral matrix equation is called the *Markov renewal equation* for $\mathbf{R}$.

To obtain the corresponding matrix integral equation for the matrix

$$\mathbf{H} = \left[H_{ij}\right], \quad [4.123]$$

we know, from relation [4.110], that

$$\mathbf{R}(t) = \mathbf{I}(t) + \mathbf{H}(t). \quad [4.124]$$

Inserting this expression of $\mathbf{R}(t)$ in relation [4.122], we obtain:

$$\mathbf{H}(t) = \mathbf{Q}(t) + \mathbf{Q} \bullet \mathbf{H}(t) \quad [4.125]$$

which is the Markov renewal equation for $\mathbf{H}$.

For $m = 1$, this last equation gives the traditional renewal equation [2.21] of Chapter 2 in the homogeneous case.

In fact, the Markov renewal equation [4.125] is a particular case of the matrix integral equation of the type:

$$\mathbf{f} = \mathbf{g} + \mathbf{Q} \bullet \mathbf{f}, \qquad [4.126]$$

called an integral equation of *Markov renewal type* (MRT), where

$$\mathbf{f} = (f_1, \ldots, f_m)', \mathbf{g} = (g_1, \ldots, g_m)' \qquad [4.127]$$

are two column vectors of functions having all their components in B, the set of single-variable measurable functions, bounded on finite intervals, or to B^+ if all their components are non-negative.

PROPOSITION 4.6.– The Markov integral equation of MRT,

$$\mathbf{f} = \mathbf{g} + \mathbf{Q} \bullet \mathbf{f} \qquad [4.128]$$

with **f**, **g** belonging to B^+, has the unique solution:

$$\mathbf{f} = \mathbf{R} \bullet \mathbf{g}. \qquad [4.129]$$

4.10.3. *Asymptotic behavior of an MRP*

We will give asymptotic results, first for the Markov renewal functions and then for solutions to integral equations of an MRT. To conclude, we will apply these results to transition probabilities of an SMP. A complete treatment can be found in Janssen & Manca [JAN 06].

We know that the renewal function R_{ij}, i, j belonging to I, is associated with the delayed renewal process, possibly transient, characterized by the couple (G_{ij}, G_{jj}) d.f. on $\mathbb{R}^+$.

Let us recall that μ_{ij} represents the mean, possibly infinite, of the d.f. G_{ij}.

PROPOSITION 4.7.– For all i, j of I, we have:

1) $$\lim_{t \to \infty} \frac{R_{ij}(t)}{t} = \frac{1}{\mu_{jj}}, \qquad [4.130]$$

2) $\lim_{t\to\infty} \frac{R_{ij}(t) - R_{ij}(t-\tau)}{\tau} = \frac{\tau}{\mu_{jj}}$, for every fixed τ. [4.131]

The next proposition, due to Barlow [BAR 62], is a useful complement to the last proposition as it gives a method for computing the values of the mean return times $\mu_{jj}, j \in I$, in the ergodic case.

PROPOSITION 4.8.– For an ergodic MRP, the mean return times satisfy the following linear system:

$$\mu_{ij} = \sum_{k \neq j} p_{ik}\mu_{kj} + \eta_i, i = 1,...,m. \quad [4.132]$$

In particular, for $i = j$, we have

$$\mu_{jj} = \frac{1}{\pi_j}\sum_k \pi_k \eta_k, j = 1,...,m, \quad [4.133]$$

where $\eta_i, i \in I$ are defined by relation [4.20], and where $\boldsymbol{\pi} = (\pi_1,...,\pi_m)$ is the unique stationary distribution of the embedded Markov chain.

REMARK 4.6.– In a similar manner, Barlow [BAR 62] proved that if $\mu_{ij}^{(2)}, i, j \in I$ is the second-order moment related to the d.f.G_{ij}, then:

$$\mu_{ij}^{(2)} = \eta_i^{(2)} + \sum_{k \neq j} \left(p_{ik}\mu_{ik}^{(2)} + 2b_{ik}\mu_{kj} \right) \quad [4.134]$$

and in particular for $i = j$:

$$\mu_{jj}^{(2)} = \frac{1}{\pi_j}\left(\sum_k \pi_k \eta_k^{(2)} + 2\sum_{k \neq j}\sum_l \pi_l p_{lk} b_k \mu_{kj} \right) \quad [4.135]$$

with

$$\eta_k^{(2)} = \int_{[0,\infty)} x^2 dH_k(x), k \in I, \quad [4.136]$$

provided that these quantities are finite.

4.10.4. *Asymptotic behavior of SMP*

(1) *Irreducible case*

Let us consider the SMP $(Z(t),\ t \geq 0)$ associated with the MRP of kernel $\mathbf{Q}$ and defined by relation [4.31].

Starting with $Z(0) = i$, we know that the probabilities of being in state j at time t:

$$\phi_{ij}(t) = P\left(Z(t) = j \middle| Z(0) = i\right). \qquad [4.137]$$

are solution of the integral system

$$\phi_{ij}(t) = \delta_{ij}(1 - H_i(t)) + \sum_k \int_0^t \phi_{kj}(t-y)dQ_{ik}(y),\ i,j \in I. \qquad [4.138]$$

It is also possible to express the transition probabilities of the SMP with the aid of the first passage time distributions $G_{ij}, i, j \in I$:

$$\phi_{ij}(t) = \phi_{jj} \bullet G_{ij}(t) + \delta_{ij}(1 - H_i(t)),\ i,j \in I. \qquad [4.139]$$

If we fix the value j in relation [4.120], we see that the m relations for i=1,...,m form a Markov renewal type equation (MRE) of form [4.109].

Applying proposition 4.6, we immediately obtain the following proposition.

PROPOSITION 4.9.– The matrix of transition probabilities

$$\mathbf{\Phi} = \left[\phi_{ij}\right] \qquad [4.140]$$

is given by

$$\mathbf{\Phi} = \mathbf{R} \bullet (\mathbf{I} - \mathbf{H}) \qquad [4.141]$$

with

$$\mathbf{H} = \left[\delta_{ij} H_i\right]. \qquad [4.142]$$

So, instead of relation [4.120], we can now write:

$$\phi_{ij}(t) = \int_{[0,t]} (1 - H_j(t-y))dR_{ij}(y). \qquad [4.143]$$

REMARK 4.7.– The "infinitesimal" quantity $dR_{ij}(y)$ ($=r_{ij}(y)dy$, if $r_{ij}(y)$ is the density of function R_{ij}, if it exists) represents the probability that there is a Markov renewal into state j in the time interval $(y,y+dy)$, starting at time 0 in state i.

Of course, the factor $(1 - H_j(t - y))$ represents the probability of not leaving state j before a time interval of length $t - y$.

The behavior of transition probabilities [4.30] will be given in the next proposition.

PROPOSITION 4.10.– If $Z = (Z(t), t \geq 0)$ is the SMP associated with an ergodic MRP of kernel **Q**, then:

$$\lim_{t\to\infty} \phi_{ij}(t) = \frac{\pi_j \eta_j}{\sum_k \pi_k \eta_k}, \quad i,j \in I. \qquad [4.144]$$

REMARK 4.8.–

(1) As the limit in relation [4.144] does not depend on i, proposition 4.8 establishes an *ergodic property* stating that:

$$\lim_{t\to\infty} \phi_{ij}(t) = \Pi_j,$$
$$\Pi_j = \frac{\pi_j \eta_j}{\sum_k \pi_k \eta_k}. \qquad [4.145]$$

(2) As the number m of states is finite, it is clear that $(\Pi_j, j \in I)$ is a probability distribution. Moreover, as $\pi_j > 0$ for all j, we also have

$$\Pi_j > 0, j \in I. \qquad [4.146]$$

So, asymptotically, every state is reachable with a strictly positive probability.

(3) In general, we have:

$$\lim_{n\to\infty} p_{ij}^{(n)} \neq \lim_{t\to\infty} \phi_{ij}(t) \qquad [4.147]$$

since of course

$$\pi_j \neq \Pi_j, j \in I. \qquad [4.148]$$

This shows that the limiting probabilities for the embedded Markov chain are not, in general, the same as taking limiting probabilities for the SMP and so semi-Markov modelization can reveal new results than with the Markov modelization.

From propositions 4.8 and 4.10, we immediately obtain the following corollary.

COROLLARY 4.1.– For an ergodic MRP, we have:

$$\Pi_j = \frac{\pi_j}{\mu_{jj}}. \qquad [4.149]$$

This result states that the limiting probability of being in state j for the SMP is the ratio of the mean sojourn time in state j to the mean return time of j.

This intuitive result also shows how the different return times and sojourn times have a crucial role in explaining why, for the embedded MC, these times have no influence.

(2) *Non-irreducible case*

It is often the case that stochastic models used for applications need non-irreducible MRP, as, for example, in the presence of an absorbing state, i.e. a state j such that

$$p_{jj} = 1. \qquad [4.150]$$

We will now see that the asymptotic behavior is easily deduced from the irreducible case studied above.

(3) *Uni-reducible case*

As for Markov chains, this is simply the case in which the embedded MC is uni-reducible so that there exist l ($l<m$) transient states, and so that the other m-l states form a recurrent class C.

We always suppose aperiodicity both for the embedded MC and for the considered MRP.

Let $T=\{1,\ldots,l\}$ be the set of transient states $(T=I-C)$. Then it can be proven

$$\lim_{t\to\infty}\phi_{ij}(t)=0,\ i,j\in T,\ \lim_{t\to\infty}\phi_{ij}(t)=G_{ij}(\infty)\frac{\pi_j\eta_j}{\sum_{k=l+1}^{m}\pi_k\eta_k},\quad i,j\in C, \qquad [4.151]$$

where $(\pi_{l+1},\ldots,\pi_m)$ represents the unique stationary probability distribution of the sub-Markov chain with C as state space where

Since

$$G_{ij}(\infty)=f_{ij}=f_{i,C} \qquad [4.152]$$

where $f_{i,C}$ is the probability that the system, starting in state i will be absorbed by the recurrent class C.

As there is only one essential class, we know that for all states i of I:

$$f_{i,C}=1, \qquad [4.153]$$

thus proving the following proposition.

PROPOSITION 4.11.– For any periodic uni-reducible MRP, we have:

$$\lim_{t\to\infty}\phi_{ij}(t)=\Pi'_j, j\in I \qquad [4.154]$$

where

$$\Pi'_j=\begin{cases}0,\ j\in T,\\ \dfrac{\pi_j\eta_j}{\sum_{k=l+1}^{m}\pi_k\eta_k}, j\in C.\end{cases} \qquad [4.155]$$

Here too, as the limit in [4.155] is independent of the initial state i, this result gives an *ergodic property*.

(4) *General case*

For any aperiodical MRP, there exists a unique partition of the state space I:

$$I = T \cup C_1 \cup \cdots \cup C_r,\ r < m, \qquad [4.156]$$

where T represents the set of transient states and $C_\nu, \nu = 1,\dots,r$ represents the νth essential class necessarily formed of positive recurrent.

From Chapter 3, we know that the system will finally enter one of the essential classes and will then stay in it forever. Thus, a slight modification of the last proposition leads to the next result.

PROPOSITION 4.12.– For any aperiodic MRP, we have:

$$\lim_{t\to\infty} \phi_{ij}(t) = \Pi'_{ij}, i, j \in I, \qquad [4.157]$$

with, for any $j \in C_\nu, \nu = 1,\dots,r$ *:*

$$\Pi'_{ij} = \begin{cases} \Pi'^{\nu}_j, & i \in C_\nu, \nu = 1,\dots,r, \\ 0, & i \in C_{\nu'}, \nu \neq \nu', \nu' = 1,\dots,r, \\ f_{i,C_\nu}\Pi'^{\nu}_j, & i \in T \end{cases} \qquad [4.158]$$

where $(\Pi'^{\nu}_j, j \in C_\nu)$ is the only stationary distribution of the sub-SMP with C_ν as state space, that is:

$$\Pi'_j = \frac{\pi_j \eta_j}{\sum_{k \in C_\nu} \pi_k \eta_k}, \qquad [4.159]$$

where $(\pi^{\nu}_k, k \in C_\nu)$ is the unique stationary distribution of the sub-Markov chain with C_ν as the state space and $(f_{i,C_\nu}, i \in T)$ is the unique solution of the linear system

$$y_i - \sum_{j \in T} p_{ij} y_j = \sum_{j \in C_\nu} p_{ij}, i \in T. \qquad [4.160]$$

Note that, in this proposition, the ergodic property is lost; this is due to the presence of the quantities f_{i,C_ν} in relation [4.158].

4.11. Markov renewal equations for the non-homogeneous case

4.11.1. *Entrance time*

This section will generalize the results given in the previous section studying the properties of non-homogeneous Markov renewal processes (NHMRP).

Let us consider a non-homogeneous MRP of kernel $\mathbf{Q}$ and, to avoid trivialities, we will assume that:

$$\sup_{i,j} Q_{ij}(s,s) < 1, \tag{4.161}$$

where the functions Q_{ij} are defined by relation [4.13].

If the starting state J_k at time T_k is i, let us define the r.v. $T_n(i|i)$, $n \geq k+1$, as the times (possibly infinite) of *successive returns* to state i, also called *successive entrance times* into $\{i\}$ from the time T_k .

It follows that the process $\left(\overline{T}_{k,n}(i|i), n \geq k\right)$ with:

$$\overline{T}_{k,k}(i|i) = 0 \tag{4.162}$$

is a non-homogeneous renewal process that could possibly be defective.

From now on, the r.v. $\overline{T}_{k,n}(i|i)$ will be called the *(n-k)th return time to state i from the time* $s = T_k$.

More generally, let us also fix state j, different from the state i already fixed; we can also define the *(n-k)th return* or *entrance time to state j*, but *starting from i as the starting state and the starting time* $s = T_k$. This time, possibly infinite as well, will be represented by $\left(T_n(j|i), n \geq 0\right)$, also using the convention that

$$T_k(j|i) = s. \tag{4.163}$$

Now, the sequence $\left(T_n(j|i), n \geq 0\right)$ is a delayed non-homogeneous renewal process with values in $\mathbb{R}^+$.

It is thus defined by two d.fs.: $G_{ij}(s,\cdot), s = T_k$ being that of $T_k(j|i)$ and $G_{jj}(s,\cdot)$ that of $T_{k+2}(j|i) - T_{k+1}(j|i)$, so that:

$$G_{ij}(s,t) = P\left(\overline{T}_{k,k+1}(j|i) \leq t\right),$$
$$G_{jj}(s',t) = P\left(\overline{T}_{k,n}(j|i) - \overline{T}_{k,n-1}(j|i) \leq t\right), T_{n-1} = s', n \geq k+2. \quad [4.164]$$

Of course, the d.f. $G_{jj}(s',\cdot), s = \overline{T}_{k+1,n}(j \mid j), n \geq k$ suffices to define the renewal process $\left(T_{k+1,n}(j|j), n \geq 0\right)$.

REMARK 4.9.– From the preceding definitions, we can also write that:

$$G_{ij}(s,t) = P\left(N_j(s,t) > 0 | J_k = i\right); i, j \in I, T_k = s,$$
$$P\left(T_{k+1}(j|i) = +\infty\right) = 1 - G_{ij}(s, +\infty) \quad [4.165]$$

and for the mean of the $\overline{T}_{k,n}(i|i)$, $n \geq k+1$, possibly infinite, we obtain:

$$\mu_{ij} = E\left(T_{k+1}(j|i)\right) = \int_s^\infty t dG_{ij}(s,t), \quad [4.166]$$

with the usual convention that

$$0 * (+\infty) = 0 \quad [4.167]$$

The means $\mu_{ij}, i, j \in I$ are called the *first entrance after time s* or *average return times after s*.

It is possible to show that the functions $G_{ij}, i, j \in I$ satisfy the following relationships:

$$G_{ij}(s,t) = \sum_{k=1}^{m} G_{kj}(s,\cdot) \bullet Q_{ik}(s,t) + (1 - G_{jj}(s,t) \bullet Q_{ij}(s,t), i, j \in I, t \geq 0. \quad [4.168]$$

where the symbol $\bullet$ represents the classical non-homogeneous Lebesgue-Stieltjes convolution product.

We will now define by A_{ij} and R_{ij} the associated *renewal functions*

$$\begin{aligned} A_{ij}(s,t) &= E(N_j(s,t) | J_k = i), \\ R_{ij}(s,t) &= E(N_j^{'}(s,t) | J_k = i) \end{aligned} \qquad [4.169]$$

where $N_j^{'}(s,t) = \delta_{ij} + N_j(s,t), j \in I$

so that

$$R_{ij}(s,t) = \delta_{ij} U_0(t) + A_{ij}(s,t). \qquad [4.170]$$

Using the classical results of the renewal theory of Chapter 2, we obtain:

$$\begin{aligned} R_{jj}(s,t) &= \sum_{n=k}^{+\infty} G_{jj}^{(n)}(s,t), j \in I, \\ R_{ij}(s,t) &= G_{ij} \cdot R_{jj}(s,t) \end{aligned} \qquad [4.171]$$

or equivalently, we have:

$$R_{ij}(s,t) = \delta_{ij} U_0(s,t) + G_{ij} \bullet \sum_{n=k}^{\infty} G_{jj}^{(n)}(s,t), i, j \in I. \qquad [4.172]$$

The following relations will express the renewal functions R_{ij}, $i, j \in I$ in function of the kernel $\mathbf{Q}$ instead of the m^2 functions G_{ij}.

PROPOSITION 4.13.– For every i and j of I, we have:

$$R_{ij}(s,t) = \sum_{n=k}^{\infty} Q_{ij}^{(n)}(s,t). \qquad [4.173]$$

Using matrix notation with:

$$\mathbf{R}(s) = \left[R_{ij}(s) \right], \qquad [4.174]$$

relation [4.173] takes the form:

$$\mathbf{R}(s) = \sum_{n=k}^{\infty} \mathbf{Q}^{(n)}(s). \qquad [4.175]$$

Let us now introduce the L-S transform of matrices.

For any matrix of suitable functions A_{ij} from $\mathbb{R}^+$ to $\mathbb{R}$ represented by

$$\mathbf{A}(s)=\left[A_{ij}(s)\right] \quad [4.176]$$

we will represent its L-S transform by:

$$\bar{\mathbf{A}}(s)=\left[\bar{A}_{ij}(s)\right] \quad [4.177]$$

with

$$\bar{A}_{ij}(s)=\int_0^\infty e^{-(t-s)}dA_{ij}(s,t). \quad [4.178]$$

Doing so for matrix $\mathbf{R}$, we obtain the matrix form of relation [4.166],

$$\bar{\mathbf{R}}(s)=\sum_{n=0}^{\infty}\left(\bar{\mathbf{Q}}(s)\right)^n. \quad [4.179]$$

From this last relation, a simple algebraic argument shows that, for any $s>0$, relations

$$\bar{\mathbf{R}}(s)(\mathbf{I}-\bar{\mathbf{Q}}(s))=(\mathbf{I}-\bar{\mathbf{Q}}(s)\bar{\mathbf{R}}(s)=\mathbf{I} \quad [4.180]$$

hold and so we also have:

$$\bar{\mathbf{R}}(s)=(\mathbf{I}-\bar{\mathbf{Q}}(s))^{-1}. \quad [4.181]$$

We have thus proved the following proposition.

PROPOSITION 4.14.– *The Markov renewal matrix* $\mathbf{R}$ *is given by*

$$\mathbf{R}(s)=\sum_{n=0}^{\infty}\mathbf{Q}^{(n)}(s),\ \mathbf{Q}^{(n)}(s)=\mathbf{Q}(s,s+1)\bullet\mathbf{Q}(s,s+2)\bullet\cdots\bullet\mathbf{Q}(s,s+n) \quad [4.182]$$

the series being convergent in $\mathbb{R}^+$, and the inverse existing for all positive s.

The knowledge of the Markov renewal matrix $\mathbf{R}(s)$ or its L-S transform $\bar{\mathbf{R}}(s)$ leads to useful expressions for d.f. of the first entrance times.

4.11.2. *The Markov renewal equation*

This section will extend the basic results related to the renewal equation developed in section 2.4 to the Markov renewal case.

Let us consider an MRP of kernel $\mathbf{Q}(s)$.

From relation [4.93], we obtain:

$$\begin{aligned} R_{ij}(s,t) &= \delta_{ij}U_0(s,t)+\sum_{n=1}^{\infty}Q_{ij}^{(n)}(s,t) \\ &= \delta_{ij}U_0(s,t)+(Q\bullet R)_{ij}(s,t). \end{aligned} \quad [4.183]$$

Using matrix notation with:

$$\mathbf{I}(s,t)=\left[\delta_{ij}U_0(s,t)\right], \quad [4.184]$$

relations [4.183] take the form:

$$\mathbf{R}(s,t)=\mathbf{I}(s,t)+\mathbf{Q}\bullet\mathbf{R}(s,t). \quad [4.185]$$

This integral matrix equation is called the *Markov renewal equation* for **R**.

To obtain the corresponding matrix integral equation for the matrix

$$\mathbf{H}(s)=\left[H_{ij}(s)\right], \quad [4.186]$$

we know, from relation [4.95], that

$$\mathbf{R}(s,t)=\mathbf{I}(s,t)+\mathbf{H}(s,t). \quad [4.187]$$

Inserting this expression of $\mathbf{R}(s,t)$ in relation [4.175], we obtain:

$$\mathbf{H}(s,t)=\mathbf{Q}(s,t)+\mathbf{Q}\bullet\mathbf{H}(s,t) \quad [4.188]$$

which is the *Markov renewal equation* for **H**.

For $m = 1$, this last equation gives the renewal equation of Chapter 3 in the non-homogeneous case.

In fact, the Markov renewal equation [4.175] is a particular case of the matrix integral equation of the type:

$$\mathbf{f} = \mathbf{g} + \mathbf{Q} \bullet \mathbf{f}, \tag{4.189}$$

called an integral equation of non-homogeneous *Markov renewal type* (MRT), where

$$\mathbf{f} = (f_1, \ldots, f_m)', \mathbf{g} = (g_1, \ldots, g_m)' \tag{4.190}$$

are two column vectors of functions having all their components in B, the set of single-variable measurable functions, bounded on finite intervals, or to B^+ if all their components are non-negative.

PROPOSITION 4.15.– The non-homogeneous Markov integral equation of MRT,

$$\mathbf{f}(s) = \mathbf{g}(s) + \mathbf{Q} \bullet \mathbf{f}(s) \tag{4.191}$$

with **f**(s), **g**(s) belonging to B^+, has the unique solution:

$$\mathbf{f}(s) = \mathbf{R}(s) \bullet \mathbf{g}(s). \tag{4.192}$$

5

Stochastic Calculus

This chapter presents the basic results concerning Itô's calculus, which is also called stochastic calculus, one of the main tools used in insurance and also the most important notions and results concerning diffusion processes intensively used in stochastic modeling.

5.1. Brownian motion

In the preceding chapters, we have seen that there exist a lot of particular stochastic processes and some of them were extensively studied, such as renewal processes, random walks, Markov chains, semi-Markov and Markov processes and their main extensions.

For building useful stochastic models in insurance and finance, it is necessary to introduce a specific stochastic process called *Brownian motion*, or the Brownian or Wiener process.

We will work on a basic complete filtered probability space satisfying the usual assumptions and denoted as $\left(\Omega,\Im,P,\left(\Im_t,t\in[0,\infty)\right)\right)$.

DEFINITION 5.1.– *The real stochastic process* $B=\left(B_t,t\in[0,\infty)\right)$ *will be called a Brownian motion (BM for short) with trend* μ *and variance* σ^2 *provided that:*

i) B is adapted to the basic filtration;

ii) B has independent increments i.e. that:

$$\forall s,t\ (0\leq s<t)\ :\ P\left(B_t-B_s\in A\middle|\Im_S\right)=P\left(B_t-B_s\in A\right),\quad \forall \text{ Borel set } B, \qquad [5.1]$$

iii) B has stationary increments, i.e.:

$$\forall s,t \ (0 \leq s < t): \ B_t - B_s \text{ has a normal distribution } N(\mu(t - s),\sigma^2(t - s)), \quad [5.2]$$

iv) $P(B_0 = x) = 1, \ (x \in \mathbb{R}).$ [5.3]

If moreover, we have:

$$\mu = 0, \ \sigma^2 = 1, \ x = 0, \quad [5.4]$$

then the Brownian motion is called *standard* (in short SBM).

Figures 5.1 and 5.2 show the two simulations of Brownian motion.

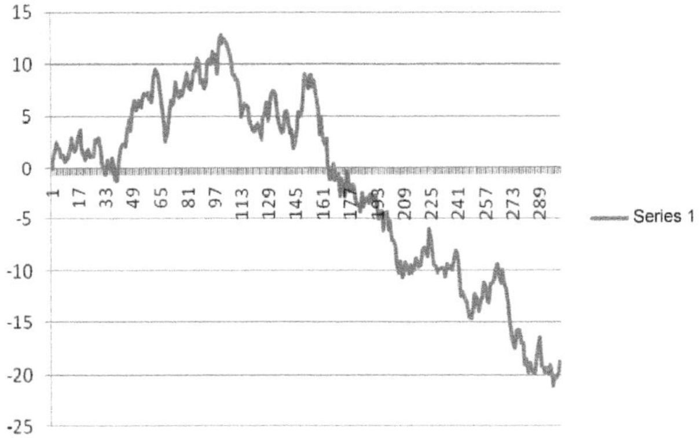

Figure 5.1. *Example 1 of a trajectory of a standard Brownian motion*

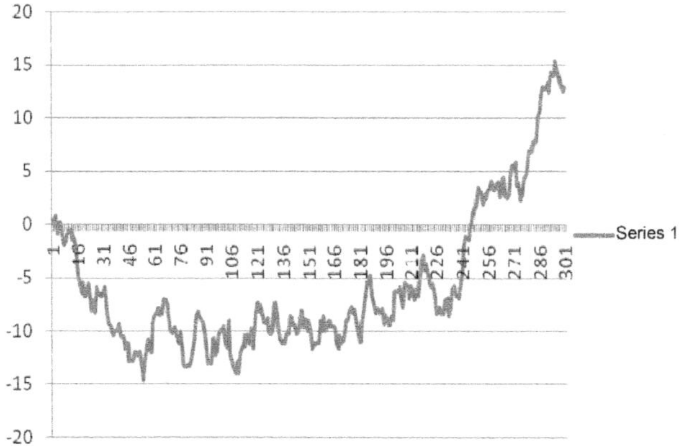

Figure 5.2. *Example 2 of a trajectory of a standard Brownian motion*

Let us now give the most important properties of the standard Brownian motion.

PROPERTY 5.1.– If B is a Brownian motion, then there exists a modification of B, the process B^*, such that B^* has a.s. continuous trajectories.

PROPERTY 5.2.– If B is a standard Brownian motion, then B is a martingale.

Indeed, from definition 5.1, we can write that:

$$\forall s,t\ (0 \le s < t)\ :\ \mathrm{E}\left[B_t - B_s \middle| \Im_S\right] = E\left[B_t - B_s\right]$$
$$0.$$

Consequently, we obtain:

$$\forall s < t : E\left[B_t \middle| B_s\right] = B_s,$$

a relation proving well that the SMB is a martingale.

Similarly, it can be proved that:

PROPERTY 5.3.– If B is a standard Brownian motion, then the process Q where:

$$Q = \left(B_t^2 - t, t \in [0,\infty)\right)$$

is a martingale.

REMARK 5.1.– It can also be proved that both properties 5.2 and 5.3 characterize a standard Brownian motion.

PROPERTY 5.4.– If B is a standard Brownian motion, then for almost all ω, the trajectory $\omega \mapsto B_t(\omega)$ is not of bounded variation on every closed interval $[a,b]$.

This explains why it is necessary for models in finance and insurance to define a new type of integral, called the Itô or stochastic integral, if we want to integrate with respect to B (see, for example, [PRO 90]).

5.2. General definition of the stochastic integral

5.2.1. *Problem of stochastic integration*

Let us consider a standard Brownian motion, $W = (W(t), t \ge 0)$ and define the following formal integral:

$$\int_0^t W(s,\omega)dW(s,\omega) \tag{5.5}$$

Though the trajectories of a Brownian motion are a.s. continuous; they are not of bounded variation on any interval $[0,t]$ and therefore, the classical integration theory of mathematical analysis cannot be applied to give the definition of the formal integral [5.5].

This is why it is necessary to construct a new theory of integration called *the stochastic* or *Itô integration*.

In particular, we will see that in this new theory, the "natural" result in classical analysis:

$$\int_0^t W(s,\omega)dW(s,\omega); \tag{5.6}$$

is here false.

More generally, the classical formula of derivation and differentiation will no more be systematically true.

5.2.2. *Stochastic integration of simple predictable processes and semi-martingales*

5.2.2.1. *Case of simple predictable processes*

Let $(\Omega,\Im,(\Im_t,t\geq 0),P)$ be a filtered complete probability space, T be a stopping time and $\Im_T$ be the σ-algebra of all the events anterior to T and introduce the following definitions.

DEFINITION 5.2.– *A stochastic process* $H=(H_t,t\geq 0)$ *is simple predictable if:*

$$\begin{aligned}&\text{i) } H_t = H_{-1}1_{\{0\}}(t)+\sum_{i=0}^{n} H_i 1_{(T_i,T_{i+1}]},\\ &\text{ii) } T_0 = 0,(T_i, i=1,...,n) \text{ is an increasing sequence of a.s.}\\ &\text{finite stopping times,}\\ &\text{iii) } \forall i=1,...,n\text{: the r.v. } H_i \text{ are } \Im_{t_i}\text{-measurable and a.s. finite.}\end{aligned} \tag{5.7}$$

DEFINITION 5.3.– *On the filtered probabilty space* $(\Omega, \Im, (\Im_t, t \geq 0), P)$, *the set of all simple predictable stochastic processes is called S and is denoted by* S_u *if it is topologized with the uniform convergence in* (t, ω).

DEFINITION 5.4.– *Given a process X is called an integrator process, the stochastic integral process denoted by* $\int_0^\infty HdX$ *of a simple predictable process H, represented by:*

$$\int_0^\infty HdX = \int_0^\infty H_s(\omega)dX_s(\omega) \tag{5.8}$$

is defined as the *operator* $I_X : S \to L^0$ where:

$$I_X(H) = H_{-1}1\{0\} + \sum_{i=0}^{n} H_i(X_{T_{i+1}} - X_{T_i}). \tag{5.9}$$

where L^0 represents the set of all r.v. with the convergence in probability provided that this operator has the following properties:

i) I_X is linear:

$$H_1, H_2 \in S : \int_0^\infty (H_1 + H_2)dX = \int_0^\infty H_1 dX + \int_0^\infty H_2 dX, \tag{5.10}$$

ii) I_X is continued:

$$(H_n) \overset{c.u.}{\rightarrow} H \Rightarrow \int_0^\infty H_n dX \overset{c.pr.}{\rightarrow} \int_0^\infty HdX. \tag{5.11}$$

We see that the continuity property relates well to two modes: the uniform convergence on *S* and the convergence in probability on L^0 (see Chapter 1).

The new problem is now to see what the "good" integrator processes are so that this definition has a meaning and satisfies properties [5.10] and [5.11].

From definition 5.4, it is clear that the linearity property is always fulfilled for simple predictable processes. To obtain classes of integrator processes that can be used, it suffices that we keep property [5.11] which the introduction of a large class of stochastic processes called *semi-martingales*.

5.2.2.2. Semi-martingales

DEFINITION 5.5.– *The stochastic process X is a total semi-martingale if:*

i) *X is càdlàg;*

ii) *X is adapted;*

iii) *the operator* $I_X : S \to L^0$ *is continuous.*

For the restriction of the integration on the interval [*0,t*], we have to give the following definition:

DEFINITION 5.6.– *The stochastic process X is a total semi-martingale if for all* $t \in [0,\infty)$, *the stopped process at t* X^t *defined by:*

$$X_s^t = \begin{cases} X_s, s < t, \\ X_t, s \geq t. \end{cases} \qquad [5.12]$$

is a total semi-martingale.

Protter [PRO 90] proved the following results.

PROPOSITION 5.1.– [PRO 90]

i) Every adapted càdlàg process of bounded variation on all compacts is a semi-martingale.

ii) Every càdlàg square integrable martingale g is a semi-martingale.

iii) Every standard Brownian motion is a semi-martingale.

PROPOSITION 5.2.– [PRO 90] If the integrator process *X* is a semi-martingale, then conditions [5.10] and [5.11] are satisfied.

In particular, from proposition 5.1, the Brownian motion can be taken as an integrator process.

5.2.3. General definition of the stochastic integral

Let us now go to the last step of stochastic integration that is to define this concept for more general processes than the simple predictable processes. It is necessary to introduce a class of stochastic processes which we can obtain by using an adequate convergence or using a technique similar to the construction of real numbers from the rational numbers or the construction of the integral of measurable functions starting from the integral of simple functions.

The basic idea, fully developed in [PRO 90], is always the same. Firstly, we define a larger class of integrable functions on which the initial class is *dkense*. Secondly, we approach each element of the new class with a sequence of elements of the initial class but using an *adequate* mode of convergence, i.e. the punctual convergence in number theory, the uniform convergence in classical integration and here the *uniform convergence in probability on every compact set.*

DEFINITION 5.7.– (*Uniform convergence in probability on every compact set*)

A sequence of stochastic processes $\left(H^n, n \geq 1\right)$ where $H^n = \left(H_t^n, t > 0\right)$ converges uniformly in probability on the compacts toward the process $H = \left(H_t, t \geq 0\right)$ if, for all t>0, we have:

$$\sup_{0 \leq s \leq t} \left| H_n^s - H_s \right| \xrightarrow{pr} 0. \qquad [5.13]$$

So, now we have four basic spaces of topologized stochastic processes:

– D: space of càdlàg simple adapted processes;

– L: space of adapted càglàd processes;

– S_u : space of simple predictable processes with the uniform convergence;

– L^0 : space of finite random variables with the convergence in probability.

The spaces of stochastic processes *D*, *L* and *S* with the *uniform* convergence in probability on the compacts are denoted, respectively, as $D_{ucp}, L_{ucp}, S_{ucp}$.

Now, we have the following result:

PROPOSITION 5.3.– [PRO 90] With the uniform convergence in probability on the compacts, the space S of simple predictable processes is dense on L.

This result leads to the extension of the definition of stochastic integral from *S* to *L*.

First, let us recall that the application:

$$I_X : S_u \mapsto L^0 \qquad [5.14]$$

defined from relation [5.2] is written under the following form:

$$I_X(H) = \int_0^\infty H_s dX_s \qquad [5.15]$$

and with the stopped process X^t :

$$I_{X^t}(H) = \int_0^t H_s dX_s \qquad [5.16]$$

For a given stochastic process *H,* this last relation defines a new stochastic process

$$J_X : J_X(H)_t = I_{X^t}(H) \qquad [5.17]$$

such that for each process $H = (H_t, t \geq 0)$, the corresponding associate process is:

$$\left(\int_0^t H_s dX_s, t \geq 0\right)$$

and so

$$J_X(H))_t = I_{X^t}(H). \qquad [5.18]$$

Protter [PRO 90] proved the following two propositions.

PROPOSITION 5.4.– If the process X is a semi-martingale, then the application:

$$J_X : S_{ucp} \mapsto D_{ucp} \qquad [5.19]$$

is continuous.

PROPOSITION 5.5.– The continuous linear operator $J_X : S_{ucp} \mapsto D_{ucp}$ can be extended to a continuous linear operator $J_X : L_{ucp} \mapsto D_{ucp}$.

This last proposition is a special case of the fundamental result that every linear operator on a subvector space can be extended in a unique way to all the vector spaces.

DEFINITION 5.8.– *If X is a semi-martingale, then continuous linear application*:

$$J_X : L_{ucp} \mapsto D_{ucp} \qquad [5.20]$$

is called a stochastic integral.

Of course, we will use the same notations as for simple processes:

$$I_X(H) = \int_0^\infty H_s dX_s \qquad [5.21]$$

$$I_{X^t}(H) = \int_0^t H_s dX_s, \ (J_X(H))_t = I_{X^t}(H) \qquad [5.22]$$

The main conclusion is thus that it is possible to define the stochastic integral on [0,t] for every adapted càglàd process as integrand process and for every semi-martingale integrator process.

EXAMPLE 5.1.– Let us consider a standard Brownian motion $B = (B_t \geq 0)$ on the filtered probability space$(\Omega, \Im, (\Im_t, t \geq 0), P)$.

From proposition 5.1, the process B is a semi-martingale and moreover continuous. Consequently, the following stochastic integral $\int_0^t B_s dB_s$ exists.

To compute its value, let us introduce the following sequence of nested partitions $(\Pi_n, n \geq 1)$ of *[0,t]* such that the sequence of these norms $(\nu_n, n \geq 1)$ tends to 0.

For every partition Π_n, we introduce the following simple function B^n defined as follows:

$$B_s^n = \sum_{k=0}^{n-1} B_{t_k} 1_{(t_k, t_{k+1}]}, \qquad [5.23]$$

with

$$\Pi_n = (t_0, \ldots, t_k, \ldots, t_n), t_0 = 0, t_n = t. \qquad [5.24]$$

From the definition of the stochastic integral of simple functions, we get:

$$\int_0^t B_s^n dB_s = \sum_{k=0}^{n-1} B_{t_k}\left(B_{t_{k+1}} - B_{t_k}\right). \qquad [5.25]$$

Using the theorem of the approximation of every continuous function by a uniformly convergent sequence of step functions, we have on $[0,t]$:

$$B^n \overset{ucp}{\rightarrow} B \qquad [5.26]$$

and so:

$$\int_0^t B_s dB_s = \lim_{\nu_n \to 0} \sum_{k=0}^{n-1} B_{t_k}\left(B_{t_{k+1}} - B_{t_k}\right). \qquad [5.27]$$

As

$$\sum_{k=0}^{n-1} B_{t_k}\left(B_{t_{k+1}} - B_{t_k}\right) = \frac{1}{2}\sum_{k=0}^{n-1}\left\{\left(B_{t_{k+1}} + B_{t_k}\right)\left(B_{t_{k+1}} - B_{t_k}\right) - \left(B_{t_{k+1}} - B_{t_k}\right)\right\}^2, \qquad [5.28]$$

or still

$$\sum_{k=0}^{n-1} B_k\left(B_{k+1} - B_k\right) = \frac{1}{2}B_t^2 - \frac{1}{2}\sum_{k=0}^{n-1}\left(B_{k+1} - B_k\right)^2, \qquad [5.29]$$

we get:

$$\int_0^t B_s dB_s = \frac{1}{2}B_t^2 - \frac{1}{2}\lim_{\nu_n \to 0}\sum_{k=0}^{n-1}(B_{t_{k+1}} - B_{t_k})^2 \qquad [5.30]$$

The final result comes from the application of the next proposition showing that the second term of this last relation tends toward $t/2$ and so:

$$\int_0^t B_s dB_s = \frac{1}{2}B_t^2 - \frac{1}{2}t. \qquad [5.31]$$

This last result illustrates the fact that the classical formulas of differential analysis are, in general, no longer true for the Itô calculus; here, in result [5.31], there is a supplementary term $-t/2$ with respect to the classical – but here incorrect – formula [5.6], called the *drift*.

Let us now prove the announced result.

PROPOSITION 5.6.– If B is a standard Brownian motion, we have, with the convergence in probability that:

$$\lim_{\nu_n \to 0} \sum_{k=0}^{n-1} (B_{t_{k+1}} - B_{t_{k+1}})^2 = t. \qquad [5.32]$$

PROOF.– With

$$\Pi_n = (t_0, \ldots, t_k, \ldots, t_n), t_0 = 0, t_n = t,$$

let us define:

$$\sum_{k=0}^{n-1} (B_{t_{k+1}} - B_{t_k})^2 = S_{n.}. \qquad [5.33]$$

From the identity:

$$S_n - t = \sum_{k=0}^{n-1} \left[(B_{t_{k+1}} - B_{t_{k+1}})^2 - (t_{k+1} - t_k) \right], \qquad [5.34]$$

we get:

$$E[S_n - t]^2 = E\left[\left(\sum_{k=0}^{n-1} \left[(B_{t_{k+1}} - B_{t_{k+1}})^2 - (t_{k+1} - t_k) \right] \right)^2 \right], \qquad [5.35]$$

and using the property that a standard Brownian motion has independent increments (see definition 5.1):

$$E[(S_n - t)^2] = E\left[\sum_{k=0}^{n-1} \left[(B_{t_{k+1}} - B_{t_{k+1}})^2 - (t_{k+1} - t_k) \right]^2 \right]. \qquad [5.36]$$

Consequently, it follows that:

$$E[(S_n - t)^2] = E\left[\sum_{k=0}^{n-1} \left[\left(\frac{B_{t_{k+1}} - B_{t_{k+1}}}{\sqrt{t_{k+1} - t_k}} \right)^2 - 1 \right]^2 (t_{k+1} - t_k)^2 \right]. \qquad [5.37]$$

Now, introduce the r.v.

$$Y_k = \frac{B_{t_{k+1}} - B_{t_k}}{\sqrt{t_{k+1} - t_k}} \tag{5.38}$$

having from definition 5.1 a $N(0,1)$ distribution so that we can write relation [5.37] under the form:

$$\begin{aligned} E[(S_n - t)^2] &= E\left[\sum_{k=0}^{n-1} \left[Y_k^2 - 1\right]^2 (t_{k+1} - t_k)^2\right], \\ &= \sum_{k=0}^{n-1} E\left[Y_k^2 - 1\right]^2 (t_{k+1} - t_k)^2. \end{aligned} \tag{5.39}$$

As the r.v. Y_k has the same distribution, we also get:

$$E[(S_n - t)^2] = E\left[(Y_1^2 - 1)^2\right] \sum_{k=0}^{n-1} (t_{k+1} - t_k)^2. \tag{5.40}$$

From the following inequality:

$$\sum_{k=0}^{n-1} \left(t_{k+1} - t_k\right)^2 \leq (b - a)\nu_n$$

we obtain:

$$E[(S_n - t)^2] \leq E\left[(Y_1^2 - 1)^2\right](b - a)\nu_n,$$

and so the result for $\nu_n \to 0$. □

REMARK 5.2.– The last proposition also shows that effectively the trajectories of a standard Brownian motion are not a.s. of bounded variation on any compact of the real set.

Indeed, from the almost surely continuity of the trajectories on $[0,t]$, there exists on this interval a subdivision $\Pi_n = \left(t_0, \ldots, t_k, \ldots, t_n\right), t_0 = 0, t_n = t$ of sufficiently small norm such that:

$$\left|B_{t_{k+1}} - B_{t_{k+1}}\right| < 1, \forall k = 0, \ldots, n-1 \tag{5.41}$$

and so:

$$\sum_{k=0}^{n-1}(B_{t_{k+1}}-B_{t_k})^2 \leq \sup_k \left|B_{t_{k+1}}-B_{t_k}\right| \sum_{k=0}^{n-1}\left|B_{t_{k+1}}-B_{t_k}\right| \qquad [5.42]$$

This last relation proves that, if the trajectories of a standard Brownian motion were a.s. of bounded variation on $[0,t]$, then the first member will tend to 0, which is in contradiction with proposition 5.7.

5.3. Itô's formula

The fact that the rules of the classical differential calculus are no longer true for the stochastic calculus implies to find a new tool of differentiation and integration. This tool was created by Itô [ITÔ 44] who proved a lemma called *Itô's lemma* whose the main result is a formula called the *Itô's formula.*

This formula became a very important basic tool for stochastic calculus and particularly in stochastic modeling.

5.3.1. *Quadratic variation of a semi-martingale*

Let us recall that we use the following notations:

$$\begin{aligned}&\int_0^t H_s dX_s = \int_{[0,t]} H_s dX_s,\\&\int_{0+}^t H_s dX_s = \int_{(0,t]} H_s dX_s\end{aligned} \qquad [5.43]$$

and so:

$$\int_0^t H_s dX_s = H_0 X_0 + \int_{0+}^t H_s dX_s. \qquad [5.44]$$

DEFINITION 5.9.– *Let X and Y be two semi-martingales, then*:

i) *the quadratic variation of X or bracket of X denoted as*:

$$[X,X] = \left([X,X]_t, t \geq 0\right) \qquad [5.45]$$

is the stochastic process

$$[X,X]_t = X_t^2 - 2\int_0^t X_{t\_}\, dX_s, \qquad (X_{0\_} = 0), \tag{5.46}$$

ii) *the quadratic covariation process of X and Y or bracket of X and Y is the stochastic process denoted by*:

$$[X,Y] = ([X,Y]_t, t \geq 0) \tag{5.47}$$

where

$$[X,Y]_t = X_t Y_t - \int_0^t X_{s-}\, dY_s - \int_0^t Y_{s-}\, dX_s. \tag{5.48}$$

REMARK 5.3.– From relations [5.46] and [5.31], it follows that for every standard Brownian motion:

$$[B,B]_t = t. \tag{5.49}$$

Protter [PRO 90] proved some interesting properties of these new processes and the most important for us are presented in the next proposition.

PROPOSITION 5.7.–

i) The process [X,X] is càdlàg, non-decreasing and adapted.

ii) The process [X,Y] is càdlàg, t bilinear and symmetric and:

$$[X,Y]_t = \frac{1}{2}\left([X+Y, X+Y]_t - [X,X]_t - [Y,Y]_t\right), \tag{5.50}$$

iii) For every sequence of partitions of stopping times:

$$T_0^n = 0, T_1^n, \ldots, T_k^n, \ldots, T_n^n = t, \tag{5.51}$$

of norms tending a.s. to 0, then:

$$X_0^2 + \sum_{k=0}^{n-1}\left(X^{T_{k+1}^n} - X^{T_k^n}\right)^2 \underset{ucp}{\rightarrow} [X,X]. \tag{5.52}$$

iv) X and Y being two semi-martingales, so is the process *[X,X]*.

v) The integration by parts formula asserts that:

$$X_tY_t = \int_0^t X_{s-}dY_s + \int_0^t Y_{s-}dX_s + [X,Y]_t, \quad [5.53]$$

vi) If it is a process of class D, then the jump process of Y, denoted as $\Delta Y = (\Delta Y_t, t \geq 0)$, is defined as:

$$\Delta Y_t = Y_t - Y_{t-}. \quad [5.54]$$

Then, for X=Y, we have:

$$\Delta[X,X]_t = (\Delta X_t)^2, \quad [5.55]$$

it follows the non-decreasing property of *[X, X]* and its decomposition in

$$[X,X]_t = [X,X]_t^c + \sum_{0\leq s\leq t}(\Delta X_s)^2,$$

or [5.56]

$$[X,X]_t = [X,X]_t^c + X_0^2 + \sum_{0<s\leq t}(\Delta X_s)^2,$$

the first term representing the "continuous" part of *[X, X]*.

5.3.2. *Itô's formula*

In classical differential calculus, it is well known that the so-called *fundamental theorem* asserts that for any integrable *function f on* [0,*t*], we have:

$$f(x) - f(x_0) = \int_{x_0}^x f'(t)dt. \quad [5.57]$$

From stochastic calculus, the problem becomes the following one: with a semi-martingale process X as integrator process, which is the additional term, if it exists, such that we can extend this result to get the following extension:

$$f(X_t) - f(X_0) = \int_{0+}^t f'(X_{s-})dX_s + \ldots. \quad [5.58]$$

For any *function f* of class $C_{\mathbb{R}}^2$, the reply is given by the next two propositions.

PROPOSITION 5.8.–

i) General Itô formula: if X is a semi-martingale and f a function of class $C^2_{\mathbb{R}}$, then the composed process $f(X)=(f(X_t),t\geq 0)$ is also a semi-martingale and moreover:

$$\begin{aligned} f(X_t)-f(X_0)=&\int_{0+}^{t} f'(X_{s-})dX_s+\frac{1}{2}\int_{0+}^{t} f'(X_{s-})d[X,X]_s^c \\ &+\sum_{0<s\leq t}\{f(X_s)-f(X_{s-})-f'(X_s)\Delta X_s\}. \end{aligned} \quad [5.59]$$

ii) Continuous case: if X is a continuous semi-martingale and f a function of class $C^2_{\mathbb{R}}$, then the composed process $f(X)=(f(X_t),t\geq 0)$ is a semi-martingale and moreover:

$$f(X_t)-f(X_0)=\int_{0+}^{t} f'(X_{s-})dX_s+\frac{1}{2}\int_{0+}^{t} f'(X_{s-})d[X,X]_s. \quad [5.60]$$

PROOF.– The last result is a direct consequence of result [5.58] as the continuity assumption on X implies that:

$$\forall s\geq 0: X_s=X_{s-},\Delta X_s=0$$

REMARK 5.4.– It is possible to show that (see [PRO 90]) the first supplementary term in the general Itô formula is nothing else but:

$$\frac{1}{2}[f'(X),X]_t^c \quad [5.61]$$

and therefore we can put Itô's formula under the form:

$$\begin{aligned} f(X_t)-f(X_0)=&\int_{0+}^{t} f'(X_{s-})dX_s+\frac{1}{2}[f'(X),X]_t^c \\ &+\sum_{0<s\leq t}\{f(X_s)-f(X_{s-})-f'(X_s)\Delta X_s\}. \end{aligned} \quad [5.62]$$

5.4. Stochastic integral with standard Brownian motion as an integrator process

Let us return to the case of a standard Brownian motion as an integrator process to particularize the general preceding results to this special case in order to obtain more precise results.

5.4.1. *Case of simple predictable processes*

On the probability space $(\Omega, \Im, (\Im_t, t \geq 0), P)$, let us consider:

– a simple predictable process defined on $[0,t]$:

$$H_s = H_k, t_k < s \leq t_{k+1}, \text{k=0,...,n-1}, \qquad [5.63]$$

$(t_0 = 0, t_1, ..., t_n = t)$ being a partition of $[0,t]$;

– a standard Brownian motion B.

From the construction of the stochastic integral, we know that:

$$\int_0^t H_s dB_s = \sum_{k=0}^{n-1} H_k (B_{t_{k+1}} - B_{t_k}). \qquad [5.64]$$

Consequently, the mean and variance of the stochastic integral are given by:

i) *Mean*:

$$E\left[\int_0^t H_s dB_s\right] = \sum_{k=0}^{n-1} E\left[H_k \left(B_{t_{k+1}} - B_{t_k}\right)\right], \qquad [5.65]$$

and as the process H is adapted and B with independent increments, we get:

$$E\left[\int_0^t H_s dB_s\right] = \sum_{k=0}^{n-1} E[H_k] E\left[B_{t_{k+1}} - B_{t_k}\right] \qquad [5.66]$$

and finally:

$$E\left[\int_0^t H_s dB_s\right] = 0. \qquad [5.67]$$

ii) *Variance*

From result [5.67]:

$$\text{var}\left(\int_0^t H_s dB_s\right) = E\left[\left(\sum_{k=0}^{n-1} H_k (B_{t_{k+1}} - B_{t_k})\right)^2\right], \qquad [5.68]$$

we get:

$$\operatorname{var}\left(\int_0^t H_s dB_s\right) = E\left[\left(\sum_{k=0}^{n-1}\sum_{l=0}^{n-1} H_k H_l (B_{t_{k+1}} - B_{t_k})(B_{t_{l+1}} - B_{t_l})\right)\right], \qquad [5.69]$$

or:

$$\begin{aligned}\operatorname{var}\left(\int_0^t H_s dB_s\right) &= \sum_{k=0}^{n-1} E\left[H_k^2 (B_{t_{k+1}} - B_{t_k})^2\right] \\ &\quad + 2E\left[\left(\sum_{k<l} H_k H_l (B_{t_{k+1}} - B_{t_k})(B_{t_{l+1}} - B_{t_l})\right)\right],\end{aligned} \qquad [5.70]$$

using the "smoothing property" from Chapter 1, we get:

$$E\left[H_k^2 (B_{t_{k+1}} - B_{t_k})^2\right] = E\left[H_k^2 (B_{t_{k+1}} - B_{t_k})^2 \middle| \Im_{t_k}\right], \qquad [5.71]$$

and thus form the fact that H is adapted to the given filtration and B with independent increments such that:

$$E\left[B_{t_{k+1}} - B_{t_k}\right] = t_{k+1} - t_k, \qquad [5.72]$$

we get:

$$E\left[H_k^2 (B_{t_{k+1}} - B_{t_k})^2\right] = E\left[H_k^2\right](t_{k+1} - t_k),\ k = 0,\ldots,n-1. \qquad [5.73]$$

With an analog reasoning, we also have that all the double products in relation [5.70] have a null expectation so that finally:

$$\operatorname{var}\left(\int_0^t H_s dB_s\right) = \sum_{k=0}^{n-1} E\left[H_k^2\right](t_{k+1} - t_k). \qquad [5.74]$$

To summarize, we have the following basic results:

$$\begin{aligned}&E\left[\int_0^t H_s dB_s\right] = \int_0^t H_s dE[B_s] = 0, \\ &\operatorname{var}\left[\int_0^t H_s dB_s\right] = E\left(\int_0^t H_s dB_s\right)^2 = E\left(\int_0^t H_s^2 ds\right) = \int_0^t E\left[H_s^2\right] ds.\end{aligned} \qquad [5.75]$$

Similarly, we can prove the following proposition.

PROPOSITION 5.9.– Under the above assumptions and if moreover the process H is square integrable, then the following process:

$$\left(\int_0^t H_s dB_s, t \geq 0\right) \quad [5.76]$$

is a square integrable $(\Im_t)$-martingale with a.s. continuous trajectories and moreover the process

$$\left(\left(\int_0^t H_s dB_s\right)^2 - \int_0^t H_s^2 ds, t \geq 0\right) \quad [5.77]$$

is a $(\Im_t)$-martingale with a.s. continuous trajectories.

Let us also mention the following property: *if X and Y are two simple predictable square integrable processes, then:*

$$E\left[\int_0^t X_s dB_s \int_0^t Y_s dB_s\right] = E\left[\int_0^t X_s Y_s ds\right] = \int_0^t E[X_s Y_s] ds. \quad [5.78]$$

5.4.2. *Extension to general integrator processes*

From what we know from the preceding section, we will use the uniform convergence in probability to extend the preceding results to the class D of adapted càdlàg processes but which is *now square integrable.*

For a process X belonging to B, there exists a process X, there exists a sequence adapted simple square integrable processes $(H_n, n \geq 0)$ *ucp converging*, to X such that in particular:

$$\int_0^t X_s dB_s = \lim_{L^2} \int_0^t H_s^n dB_s. \quad [5.79]$$

From this result, it follows that all the properties of the preceding section 5.1 remain true in this general case.

5.5. Stochastic differentiation

5.5.1. *Stochastic differential*

On the probability space $(\Omega, \Im, (\Im_t, t \geq 0), P)$, let us consider an adapted standard Brownian motion B and two adapted processes a and b sufficiently smooth.

DEFINITION 5.10.– The *stochastic process*:

$$\xi = (\xi(t), t \geq 0) \qquad [5.80]$$

has as stochastic differential on [0,T]

$$d\xi(t) = a(t)dt + b(t)dB(t) \qquad [5.81]$$

if and only if:

$$\begin{aligned} &\forall t_1, t_2 : 0 \leq t_1 < t_2 \leq T : \\ &\xi(t_2) - \xi(t_1) = \int_{t_1}^{t_2} a(t)dt + \int_{t_1}^{t_2} b(t)dB(t). \end{aligned} \qquad [5.82]$$

5.5.2. *Particular cases*

1) Result [5.31] gives:

$$\int_0^t B_s dB_s = \frac{1}{2} B_t^2 - \frac{1}{2} t. \qquad [5.83]$$

Consequently, we also have:

$$\int_{t_1}^{t_2} B_s dB_s = \frac{1}{2}(B_{t_2}^2 - B_{t_1}^2) - \frac{1}{2}(t_2 - t_1) \qquad [5.84]$$

and from our new definition 5.10, it follows that:

$$dB_t^2 = dt + 2B_t dB_t. \qquad [5.85]$$

2) From the definition of the stochastic integral, we know that:

$$\int_{t_1}^{t_2} tdB_t = \lim_n \sum_{k=1}^{n-1} t_{n,k} \left[B_{t_{n,k+1}} - B_{t_{n,k}} \right], \tag{5.86}$$

$(t_{n,1} = t_1, \ldots, t_{n,k}, \ldots, t_{n,n} = t_2)$ being a subdivision of order n of the interval $[t_1, t_2]$.

Moreover, from the definition of the classical Lebesgue integral, we get:

$$\int_{t_1}^{t_2} B_t dt = \lim_n \sum_{k=0}^{n-1} B_{t_{n,k+1}} (t_{n,k+1} - t_{n,k}). \tag{5.87}$$

Adding member-to-member relations [5.86] and [5.87], we get:

$$\int_{t_1}^{t_2} B_t dt + \int_{t_1}^{t_2} tdB_t + = \lim_n \sum_{k=1}^{n-1} \left[t_{n,k+1} B_{t_{n,k+1}} - t_{n,k} B_{t_{n,k}} \right] \tag{5.88}$$

and so:

$$\int_{t_1}^{t_2} B_t dt + \int_{t_1}^{t_2} tdB_t = t_2 B_{t_2} - t_1 B_{t_1} \tag{5.89}$$

or in terms of stochastic differential:

$$d(tB_t) = B_t dt + tdB_t, \tag{5.90}$$

the formula is also different from the one of the classical calculus.

5.5.3. *Other forms of Itô's formula*

Now using the concept of stochastic differential, we will give other forms of Itô's formula.

5.5.3.1. *Stochastic differential of a product*

On the probability space $(\Omega, \Im, (\Im_t, t \geq 0), P)$, let us consider an adapted standard Brownian motion B and four càdlàg adapted processes $a_i, b_i, i = 1,2$ of class D and sufficiently smooth defining the following two stochastic differentials:

$$d\xi_i(t) = a_i(t)dt + b_i(t)dB(t), i = 1,2. \tag{5.91}$$

Then, we have as next result:

PROPOSITION 5.10.– [FRI 75] *The process* $\xi_1\xi_2$ *is differentiable (in Itô'sense) and:*

$$d\left(\xi_1(t)\xi_2(t)\right) = \xi_1(t)d\xi_2(t) + \xi_2(t)d\xi_1(t) + b_1(t)b_2(t)dt. \quad [5.92]$$

5.5.3.2. *Examples*

1) With $\xi_1(t) = \xi_2(t) = B(t)$, we find this known result (see relation [5.85]):

$$d\left(B^2(t)\right) = 2B(t)dB(t) + dt\,.$$

2) Similarly, we can find [5.89] concerning

$$d(tB(t)) = tdB(t) + B(t)dt,$$

with

$$\begin{aligned} &\xi_1(t) = t \Rightarrow a_1(t) = 1, b_1(t) = 0, \\ &\xi_2(t) = B(t) \Rightarrow a_1(t) = 0, b_1(t) = 1. \end{aligned} \quad [5.93]$$

5.5.3.3. *Itô's formula with time dependence*

For our applications, the main result is the so-called *Itô lemma* or the *Itô formula*, which is equivalent to the rule of derivatives for composed functions in the classical differential calculus, but now with a function f of two variables.

Starting with:

$$d\xi(t) = a(t)dt + b(t)dB(t), \quad [5.94]$$

let f be a function of two non-negative real variables x, t such that:

$$f \in C^0_{\mathbb{R}\times\mathbb{R}^+}, f_x, f_{xx}, f_t \in C^0_{\mathbb{R}\times\mathbb{R}^+}. \quad [5.95]$$

Then, the composed stochastic process:

$$\left(f(\xi(t),t), t \geq 0\right) \quad [5.96]$$

is also Itô's differentiable and its stochastic differential is given by:

$$d\big(f(\xi(t),t)\big)=$$
$$\left[\frac{\partial f}{\partial x}(\xi(t),t)a(t)+\frac{\partial f}{\partial t}(\xi(t),t)+\frac{1}{2}\frac{\partial^2}{\partial^2 x^2}f(\xi(t),t)b^2(t)\right]dt \qquad [5.97]$$
$$+\frac{\partial f}{\partial x}(\xi(t),t)b(t)dB(t).$$

REMARK 5.5.– Compared with the classical differential calculus, we know that in this case, we should have:

$$d\big(f(\xi(t),t)\big)=\left[\frac{\partial f}{\partial x}(\xi(t),t)a(t)+\frac{\partial f}{\partial t}(\xi(t),t)\right]dt$$
$$+\frac{\partial f}{\partial x}(\xi(t),t)b(t)dB(t). \qquad [5.98]$$

Therefore, the difference between [5.96] and [5.97] is the *supplementary term:*

$$\frac{1}{2}\frac{\partial^2}{\partial^2 x^2}f(\xi(t),t)b^2(t) \qquad [5.99]$$

appearing in [5.97] and which is null *iff* in two cases:

1) f is a linear function of x;

2) b is identically equal to 0.

EXAMPLE 5.2.–

1) For ξ given by:

$$d\xi(t)=dB(t),$$
$$\xi(0)=0. \qquad [5.100]$$

Using notation [5.94], we get:

$$\text{a(t)=0, b(t)=1.} \qquad [5.101]$$

With the aid of Itô's formula, the value of $de^{B(t)}$ is thus given by:

$$de^{B(t)}=\frac{1}{2}e^{B(t)}dt+e^{B(t)}dB(t). \qquad [5.102]$$

As we can see, the first term is the supplementary term with respect to the classical formula and is called the drift.

REMARK 5.6 (INTERPRETATION OF ITÔ'S FORMULA).–

Itô's formula simply means that the composed stochastic process:

$$\left((f(\xi(t),t)-f(\xi(0),0),t\geq 0\right) \quad [5.103]$$

is stochastically equivalent to the following stochastic process:

$$\begin{aligned}&\left(\int_0^t\left[f_t(\xi(s),s)ds+f_x(\xi(s),s)a(s)+\frac{1}{2}f_{xx}(\xi(s),s)b^2(s)\right]ds\right)\\&+\int_0^t f_x(\xi(s),s)b(s)dB(s),t\geq 0).\end{aligned} \quad [5.104]$$

5.5.3.4. *Other extensions of Itô's formula*

i) Extension to a multidimensional process

It is possible to extend Itô's formula in the following way. Let $\xi=(\xi(t),t\geq 0)$ be an m-dimensional stochastic process:

$$\xi(t)=(\xi_1(t),\ldots,\xi_m(t))' \quad [5.105]$$

with every component having a stochastic differential given by:

$$d\xi_i(t)=a_i(t)dt+b_i(t)dB(t), i=1,\ldots,m\,. \quad [5.106]$$

Then, it is possible to show that the stochastic differential of the one-dimensional process:

$$\left(f(\xi(t),t),t\geq 0\right), \quad [5.107]$$

where f is a real function of m+1 variables:

$$f(\mathbf{x},t)=f(x_1,\ldots,x_m,t) \quad [5.108]$$

satisfying the following assumptions:

$$f\in C^0_{\mathbb{R}^m\times\mathbb{R}^+}, f_{x_i}, i=1,\ldots,m, f_{x_ix_j}, i,j=1,\ldots,m, f_t\in C^0_{\mathbb{R}^m\times\mathbb{R}^+} \quad [5.109]$$

exists and is given by:

$$d\left(f(\xi(t),t)\right)=\left[\sum_{i=1}^{m}\frac{\partial f}{\partial x_i}(\xi(t),t)a_i(t)+\frac{\partial f}{\partial t}(\xi(t),t)\right.$$
$$\left.+\frac{1}{2}\sum_{i,j=1}^{m}\frac{\partial^2}{\partial x_i\partial x_j}f(\xi(t),t)b_i(t)b_j(t)\right]dt+\sum_{i=1}^{m}\frac{\partial f}{\partial x_i}(\xi(t),t)b_i(t)dB(t) \quad [5.110]$$

Here, the supplementary time is given by:

$$\frac{1}{2}\sum_{i,j=1}^{m}\frac{\partial^2}{\partial x_i\partial x_j}f(\xi(t),t)b_i(t)b_j(t) \quad [5.111]$$

ii) Extension to a multidimensional process with stochastic dynamics

The second possible extension also starts with an m-dimensional stochastic process $\xi(t)=(\xi_1(t),\ldots,\xi_m(t))'$ such that its dynamics is governed by the following stochastic differential:

$$d\xi(t)=\mathbf{a}(t)dt+\mathbf{b}(t)d\mathbf{B}(t), i=1,\ldots,m \quad [5.112]$$

a being an m-dimensional random vector of class L or D and **b** a stochastic matrix mxn whose elements are stochastic processes of class L and **B** an n-vector of n independent standard Brownian motions.

As in the preceding section, we are interested in the stochastic differential of the one-dimensional process:

$$\left(f(\xi(t),t),t\geq 0\right), \quad [5.113]$$

with f being a real function of m+1 variables:

$$f(\mathbf{x},t)=f(x_1,\ldots,x_m,t) \quad [5.114]$$

satisfying the following assumptions:

$$f\in C^0_{\mathbb{R}^m\times\mathbb{R}^+}, f_{x_i}, i=1,\ldots,m, f_{x_ix_j}, i,j=1,\ldots,m, f_t\in C^0_{\mathbb{R}^m\times\mathbb{R}^+}. \quad [5.115]$$

Under these assumptions, it is still possible to *show that the composed* stochastic process $\left(f(\xi(t),t),t\geq 0\right)$ is Itô's differentiable and that its stochastic differential is given by:

$$
\begin{aligned}
&d\left(f\left(\xi(t),t\right)\right)= \\
&\left[\sum_{i=1}^{m}\frac{\partial f}{\partial x_i}\left(\xi(t),t\right)a_i(t)+\frac{\partial f}{\partial t}\left(\xi(t),t\right)+\frac{1}{2}\sum_{i,j=1}^{m}\sigma_{ij}(t)\frac{\partial^2}{\partial x_i\partial x_j}f\left(\xi(t),t\right)\right]dt \\
&+\sum_{i,j=1}^{m}\frac{\partial f}{\partial x_i}\left(\xi(t),t\right)b_{ij}(t)dB_j(t) \\
&\sigma_{ij}(t)=\frac{1}{2}\left(\mathbf{bb'}(t)\right)_{ij}
\end{aligned}
\quad [5.116]
$$

Using matrix notation, we can rewrite this last expression under the form:

$$
\begin{aligned}
&d\left(f\left(\xi(t),t\right)\right)=\frac{\partial f}{\partial t}\left(\xi(t),t\right)dt+\text{grad}f(t)d\xi(t)+\frac{1}{2}\text{tr}(\mathbf{bb}')(t)\mathbf{f}_{\mathbf{xx}}(t)dt, \\
&\mathbf{f}_{\mathbf{xx}}(t)=\left[\frac{\partial^2 f}{\partial x_i\partial x_j}(t)\right].
\end{aligned}
\quad [5.117]
$$

Here, the supplementary time is given by:

$$
\frac{1}{2}\text{tr}(\mathbf{bb}')(t)\mathbf{f}_{\mathbf{xx}}(t)dt \quad [5.118]
$$

iii) Extension to the multidimensional case with dependent BM

The last extension concerns the case of a vectorial Brownian motion **B** whose components are *m dependent* standard Brownian motions.

This means that:

$$
\forall i,j,\forall s,t(s<t):E\left[(B_i(t)-B_i(s))(B_i(t)-B_i(s))\right]=\rho_{ij}(t-s). \quad [5.119]
$$

The matrix $\mathbf{Q}=\left[\rho_{ij}\right]$ is called the *correlation matrix* of the vector Brownian motion $\mathbf{B}=\left(\mathbf{B}(t),t\geq 0\right)$.

If $\mathbf{Q}=\mathbf{I}$, the vector Brownian motion $\mathbf{B}=\left(\mathbf{B}(t),t\geq 0\right)$ is called *standard* if moreover $\mathbf{B}(0)=0$.

In the case of an *n*-dimensional Brownian motion and with the same assumptions of the function *f* as above, Itô's formula becomes:

$$d\left(f\left(\xi(t),t\right)\right)=\frac{\partial f}{\partial t}\left(\xi(t),t\right)dt+\mathrm{grad}f(t)d\xi(t)+\frac{1}{2}\mathrm{tr}(\mathbf{bQb}')(t)\mathbf{f}_{\mathbf{xx}}(t)dt,$$
$$\mathbf{f}_{\mathbf{xx}}(t)=\left[\frac{\partial^2 f}{\partial x_i\partial x_j}(t)\right]. \qquad [5.120]$$

5.6. Stochastic differential equations

5.6.1. *Existence and unicity general theorem*

The problem is, as in the deterministic case, the following one: for the stochastic differential:

$$d\xi(t)=\mu(\xi(t),t)dt+\sigma(\xi(t),t)dB(t),$$
$$\xi(0)=\xi_0, a.s. \qquad [5.121]$$

with $B=\left(B(t),t\geq 0\right)$ as standard Brownian motion on the complete filtered probability space $\left(\Omega,\Im,(\Im_t,t\geq 0),P\right)$, find, if possible, a stochastic process:

$$\xi=\left(\xi(t),t\in[0,T]\right) \qquad [5.122]$$

satisfying on the interval [0,*T*] the relations [5.121], under minimal assumptions on the two functions μ,σ from $\mathbb{R}\times[0,T]\mapsto\mathbb{R}$.

Relation [5.121] is called a *stochastic differential equation* (SDE).

Gikhman and Skorokhod [GIK 69] proved a general *theorem of existence and unicity* also given, in a more modern form, by Protter [PRO 90].

Under a relatively simple form, the main result is the following.

PROPOSITION 5.11 (GENERAL THEOREM OF EXISTENCE AND UNICITY).–

Let us consider the following SDE:

$$d\xi(t)=\mu(\xi(t),t)dt+\sigma(\xi(t),t)dB(t),$$
$$\xi(0)=\xi_0, a.s. \qquad [5.123]$$

under the following assumptions:

i) the functions μ,σ are measurable functions from $\mathbb{R}\times[0,T]\mapsto\mathbb{R}$ verifying a Lipschitz condition in the first variable:

$$\begin{aligned}&\forall(x_1,t),(x_2,t)\in\mathbb{R}\times[0,T]:\\&|\mu(x_1,t)-\mu(x_2,t)|\leq\bar{K}|x_1-x_2|,\\&|\sigma(x_1,t)-\sigma(x_2,t)|\leq\bar{K}|x_1-x_2|,\end{aligned}\qquad[5.124]$$

$\bar{K}$ being a positive constant;

ii) on $\mathbb{R}\times[0,T]$, the functions μ,σ are linearly bounded:

$$|\mu(x,t)|\leq K(1+|x|),|\sigma(x,t)|\leq K(1+|x|),\qquad[5.125]$$

K being a positive constant;

iii) the r.v. ξ_0 belongs to $L^2(\Omega,\Im,P)$ and is independent *of the* σ*-algebra* $\sigma(B(t),t\in[0,T])$,

then, there exists a solution belonging for all $t\in[0,T]$, to $L^2(\Omega,\Im,P)$, continuous and a.s. unique on $[0,T]$.

REMARK 5.7.–

1) The initial condition:

$$\xi(0)=x_0,\in R\qquad[5.126]$$

of course satisfies the assumption iii).

2) This theorem can be extended in the case of an SDE on $[s,s+T]$, with the initial condition:

$$\xi(s)=\xi_s\qquad[5.127]$$

where $\xi(s)$ is a r.v. belonging to $L^2(\Omega,\Im,P)$ and is independent of the σ-algebra $\sigma(B(s+\tau)-B(s),\tau\in[0,T])$.

3) It is also possible to prove that:

$$E\left[\sup_{[0,T]}|\xi(t)|^2\right]\leq C\left(1+E\left[\xi_0^2\right]\right), \qquad [5.128]$$

where C is a constant depending only on K and T.

In proposition 5.11, the coefficients μ,σ are deterministic functions but it is possible to extend it in the stochastic case. Then, formally, we have:

$$\mu(x,t)=\mu(x,t,\omega),\sigma(x,t)=\sigma(x,t,\omega),\forall x\in\mathbb{R},\forall t\in[0,T]. \qquad [5.129]$$

The initial condition [5.126] becomes:

$$\xi(0)=\varphi(0), \qquad [5.130]$$

where

$$\varphi=\left(\varphi(t),t\in[0,T]\right) \qquad [5.131]$$

is the given initial process.

The extension of proposition 5.12 is now given.

PROPOSITION 5.12 (CASE OF RANDOM COEFFICIENTS).–

For the SDE:

$$\begin{aligned}&d\xi(t)=d\varphi(t)+\mu(\xi(t),t)dt+\sigma(\xi(t),t)dB(t),\\&\xi(0)=\varphi(0),\end{aligned} \qquad [5.132]$$

where:

i) the processes μ,σ are measurable as functions from $\mathbb{R}\times[0,T]\times\Omega\mapsto\mathbb{R}$, adapted and Lipschitzian in the first variable, i.e. with probability 1:

$$\begin{aligned}&\forall(x_1,t),(x_2,t)\in R\times[0,T]:\\&|\mu(x_1,t)-\mu(x_2,t)|\leq\overline{K}|x_1-x_2|,\\&|\sigma(x_1,t)-\sigma(x_2,t)|\leq\overline{K}|x_1-x_2|,\end{aligned} \qquad [5.133]$$

$\overline{K}$ being a positive constant;

ii) the processes μ, σ are measurable as functions from $\mathbb{R}\times[0,T]\times\Omega\mapsto\mathbb{R}$, satisfy a.s. the following condition:

$$|\mu(x,t)|^2+|\sigma(x,t)|^2\le K^2(1+x^2), \qquad [5.134]$$

K being a positive constant;

iii) the process $\varphi=(\varphi(t), t\in[0,T])$ is of bounded variation, adapted and such that

$$E\left[\sup_{[0,T]}|\varphi(t)|^2\right]<\infty, \qquad [5.135]$$

then, there exists a solution belonging for $t\in[0,T]$, to $L^2(\Omega,\Im,P)$; moreover, if ξ_1,ξ_2 are two solutions, then they are stochastically equivalent i.e.:

$$P[\xi_1(t)=\xi_2(t)]=1, \forall t\in[0,T]. \qquad [5.136]$$

Finally, if the process φ is continuous a.s. on $[0,T]$, then there exists a.s. unicity on $[0,T]$:

$$P\left[\sup_{[0,T]}\{t:|\xi_1(t)-\xi_2(t)|\}>0\right]=0. \qquad [5.137]$$

REMARK 5.8.– This theorem can be extended in the case of an SDE on $[s,s+T]$.

The proofs of these two fundamental propositions use the method of *successive approximations used in the* deterministic case under the name of *Piccard method*: on $[0,T]$, we begin to use the following very rough approximation:

$$\xi_0(t)=\xi_0 \qquad [5.138]$$

and by induction, on constructs on $[0,T]$, the following sequence of stochastic processes $\xi_n=(\xi_n(t), n>0)$ is *defined* by:

$$\xi_{n+1}(t)=\xi_0+\int_0^t\mu(\xi_n(s),s)ds+\int_0^t\sigma(\xi_n(s),s)dB(s). \qquad [5.139]$$

Thus, it is possible to show (see, for example, [FRI 75]) that the sequence $\xi_n = (\xi_n(t), n > 0)$ converges uniformly a.s. on $[0,T]$ toward the stochastic process $\xi = (\xi(t), 0 \leq t \leq T)$, which is well a solution of the considered SDE [5.132].

Using [5.133], Friedman [FRI 75] also proved the a.s. unicity.

5.6.2. *Solution of stochastic differential equations*

5.6.2.1. *Canonical form of an SDE*

Let us consider the following general SDE:

$$\begin{aligned} &d\xi(t) = d\varphi(t) + \mu(\xi(t),t)dt + \sigma(\xi(t),t)dB(t), \\ &\xi(0) = \varphi(0), \end{aligned} \qquad [5.140]$$

where $B = (B(t), t \geq 0)$ is a standard Brownian motion *on* $(\Omega, \Im, (\Im_t, t \geq 0), P)$.

The general procedure to find the process $\xi = (\xi(t), t \in [0,T])$ solution of this SDE under the assumptions of proposition 5.11 is to try to put this SDE under its *canonical form,* that is to say:

$$\begin{aligned} &d\xi(t) = a(t)dt + b(t)dB(t), \\ &\xi(0) = \xi_0, \end{aligned} \qquad [5.141]$$

with known a and b functions or stochastic processes. If so, the unique solution of the considered SDE takes the form:

$$\xi(t) = \xi_0 + \int_0^t a(s)ds + \int_0^t b(s)dB(s). \qquad [5.142]$$

More generally, we can look for a transformation f in two variables x and t, monotone in t satisfying the assumptions of the Itô's lemma *and such* that:

$$df(\xi(t),t) = \overline{A}(t)dt + \overline{B}(t)dB(t) \qquad [5.143]$$

In this case, we obtain:

$$f(\xi(t),t) = f(\xi(0),0) + \int_0^t \overline{A}(s)ds + \int_0^t \overline{B}(s)dB(s) \qquad [5.144]$$

Then taking the inverse of the function f with respect to the first variable, we find the form of $\xi(t), t \in [0, T]$.

5.6.2.2. *Particular cases*

We will present two important particular cases useful in insurance and finance for stochastic interest rate evolution model with the Ornstein–Uhlenbeck–Vasicek and the stochastic evolution of assets with the Black–Scholes–Samuelson model.

i) *The OUV (Ornstein–Uhlenbeck–Vasicek)*

This SDE is given by:

$$\begin{aligned} dr(t) &= a(b - r(t))dt + \sigma dB(t), \\ r(0) &= r_0. \end{aligned} \quad [5.145]$$

Starting first from the deterministic part:

$$\begin{aligned} dr(t) &= a(b - r(t))dt \\ r(0) &= r_0, \end{aligned} \quad [5.146]$$

we get:

$$r(t) = b + ce^{-at}, \quad [5.147]$$

c constant.

Now, let us suppose that the solution of the OUV SDE can be written under the form:

$$r(t) = b + c(t)e^{-at}, \quad [5.148]$$

where c is an unknown stochastic process as done in the classical *method of constant variation.*

From relation [5.148], we get:

$$dr = e^{-at}dc - ae^{-at}c(t)dt. \quad [5.149]$$

And if we use [5.147] again to express c, we obtain:

$$dr = e^{-at}dc + a(b - r(t))dt. \quad [5.150]$$

The comparison with the [5.145] leads to:

$$e^{-at}dc + a(b - r(t))dt = a(b - r(t))dt + \sigma dB(t),$$
or
$$e^{-at}dc = \sigma dB(t). \qquad [5.151]$$

Therefore, we get the reduced form of the SDE for the process c under the following form:

$$dc(t) = \sigma e^{at} dB(t) \qquad [5.152]$$

and, consequently,

$$c(t) = c_0 + \sigma \int_0^t e^{as} dB(s),$$
with
$$c_0 = c(0). \qquad [5.153]$$

Returning to relation [5.147], we find the solution under the form:

$$r(t) = b + e^{-at}(c_0 + \sigma \int_0^t e^{as} dB(s). \qquad [5.154]$$

Taking $t = 0$, we get:

$$c_0 = r_0 - b.$$

The final form of the solution of the OUV SDE is given by:

$$r(t) = b + (r_0 - b)e^{-at} + \sigma e^{-at} \int_0^t e^{as} dB(s). \qquad [5.155]$$

ii) *Solution of the SDE of Black–Scholes–Samuelson*

The most famous SDE of stochastic finance was originally introduced by Samuelson [SAM 65] and was later used by Black and Scholes [BLA 73] in option theory. It has the form:

$$dC(t) = \mu C(t)dt + \sigma C(t)dB_t,$$
$$C(0) = C_0. \qquad [5.156]$$

Usually, the process C represents the value of an asset depending randomly on time.

The parameter μ is called the *trend* of this asset and the parameter σ is its *volatility* on an annual basis.

To obtain the solution of the SDE, let us use the following transform:

$$y(t) = \ln C(t). \quad [5.157]$$

Using Itô's formula, we get:

$$d\big(y(C(t),t)\big) = \left[\frac{\partial y}{\partial x}(C(t),t)a(t) + \frac{\partial y}{\partial t}(C(t),t) + \frac{1}{2}\frac{\partial^2 y}{\partial^2 x^2}(C(t),t)b^2(t)\right]dt$$
$$+\frac{\partial y}{\partial x}(C(t),t)b(t)dB(t).$$

with here:

$$\begin{aligned} &a(t) = \mu C(t), b(t) = \sigma C(t), \\ &f(x,t) = \ln x. \end{aligned} \quad [5.158]$$

After simplification, we find the reduced form for the unknown process C:

$$d\ln C(t) = (\mu - \frac{\sigma^2}{2})dt + \sigma dB(t). \quad [5.159]$$

And by integration:

$$\ln C(t) - \ln C_0 = (\mu - \frac{\sigma^2}{2})t + \sigma B(t) \quad [5.160]$$

So that we obtain the solution of the BSS SDE:

$$C(t) = C_0 e^{(\mu - \frac{\sigma^2}{2})t} e^{\sigma B(t)}. \quad [5.161]$$

5.6.3. *Diffusion processes*

5.6.3.1. *Definition and Markov property*

Let us consider again the SDE [5.140] *under* the assumptions of proposition 5.11.

The solution $\xi = (\xi(t), t \in [0,T])$, of this SDE, is called a *diffusion process or Itô process*.

Let *s* and *t* such that: $0 \le s < t \le T$ and suppose that $\xi(s) = x$.

From the theorem of existence and unicity, we know that on the interval $[s,T]$ there exists only one process, denoted as $\xi_{x,s}$ solution of the SDE [5.140] such that:

$$\xi_{x,s}(s) = x \qquad [5.162]$$

So, it is clear that, setting $x = \xi(t)$, we have the Markov property for the ξ-process in continuous time, of course generally non-homogeneous.

More precisely, we have the following propositions.

PROPOSITION 5.13.– Under the assumptions of proposition 5.11 and if, for each t, $\Im_t$ represents the σ-algebra generated be ξ_0 and the set $(B(s), s \le t))$, then the a.s. unique stochastic process solution of [5.140] is Markovian, i.e. satisfies a.s.:

$$P\left[\xi(t) \in A \middle| \Im_s\right] = P\left[\xi(t) \in A \middle| \xi(s)\right] (= p(s, \xi(s), t, A)) \qquad [5.163]$$

for all t > s and for all Borel set A.

PROPOSITION 5.14.– The function of $\mathbb{R}^+ \times \mathbb{R} \times \mathbb{R}^+ \times \beta \mapsto [0,1]$ defined by relation [5.163], satisfies the following properties:

i) for all fixed s,x,t, p(s,x,t) is a probability measure on $\mathbb{R}$;

ii) for all fixed s,t,A, p(s,t,A) is Borel-measurable;

iii) the function p satisfies the Chapman–Kolmogorov equations:

$$\forall 0 \le s < t < \tau, x \in R, A \in \beta:$$

$$\int_R p(s,x,t,dy) p(t,y,\tau,A) = p(s,x,\tau,A). \qquad [5.164]$$

iv) the process solution $\xi = (\xi(s), s \geq 0)$ is a Feller process; i.e. for all continuous bounded function of $\mathbb{R} \mapsto \mathbb{R}$, the application

$$(s,x) \mapsto \int f(y) p(s,x,s+t,dy) \qquad [5.165]$$

is continuous;

(v) the process solution $\xi = (\xi(s), s \geq 0)$ satisfies the strong Markov property, i.e. condition [5.163] but where s and t are replaced by stopping times.

REMARK 5.9.–

a) In these two propositions, the transition probability function $p(s, \xi(s), t, A)$ in general depends on s and in this case, the Markov process is called *time non-homogeneous* or simply *non-homogeneous,* terminology already used in the preceding chapters.

b) If the drift and the diffusion coefficient are continuous functions, it can be shown that:

i)

$$\forall \varepsilon > 0, t \geq 0, x \in R:$$

$$\lim_{h \downarrow 0} \frac{1}{h} \int_{|y-x|>\varepsilon} p(t,x,t+h,dy) = 0, \qquad [5.166]$$

ii)

$$\forall \varepsilon > 0, t \geq 0, x \in R:$$

$$a) \lim_{h \downarrow 0} \frac{1}{h} \int_{|y-x|<\varepsilon} (y-x) p(t,x,t+h,dy) = \mu(x,t), \qquad [5.167]$$

$$b) \lim_{h \downarrow 0} \frac{1}{h} \int_{|y-x|<\varepsilon} (y-x)^2 p(t,x,t+h,dy) = \sigma^2(x,t),$$

For the applications of such processes in insurance and finance, it is interesting to give the interpretations of these last properties:

1) The probability for the process $\xi = (\xi(s), s \geq 0)$ to have a jump of amplitude more than ε between t and $t+h$ is $o(h)$.

Consequently, the process $\xi = (\xi(s), s \geq 0)$ is continuous in probability.

2) Properties *a* and *b* can be rewritten as follows:

$$\begin{aligned} &a) E\left[\xi(t+h) - \xi(t) \middle| \xi(t) = x\right] = \mu(x,t)h + o(h), \\ &b) E\left[\left|\xi(t+h) - \xi(t)\right|^2 \middle| \xi(t) = x\right] = \sigma^2(x,t)h + o(h). \end{aligned} \quad [5.168]$$

Consequently, the drift μ gives the rate of the conditional mean of the increment of the diffusion process *on the infinitesimal time (t,t+h) interval* and the square of the diffusion coefficient of diffusion σ the conditional variance of this increment as the square of the mean is of order $o(h)$.

5.6.3.2. *The Fokker–Planck partial differential equation*

If the function p has a density p', then it is solution of *the partial differential equation of Fokker–Planck:*

$$\frac{\partial p'}{\partial t} + \frac{\partial}{\partial x}(\sigma(x,t)p') - \frac{1}{2}\frac{\partial^2}{\partial x^2}(\mu(x,t)p') = 0. \quad [5.169]$$

EXAMPLE 5.3.– For the Ornstein–Uhlenbeck–Vasicek process defined in section 5.6.3.1, it can be shown that:

$$p'(s,x,t,y) = \frac{1}{\sqrt{2\pi V_t}} e^{-\frac{1}{2V_t}(x-M_t)^2}, \quad [5.170]$$

M_t, V_t representing, respectively, the mean and variance of $\xi(t)$ given by (see [JAN 09b], Chapter 16):

$$\begin{aligned} &M_t = b + (\xi_0 - b)e^{-at}, \\ &V_t = \frac{\sigma^2}{2a}(1 - e^{-2at}). \end{aligned} \quad [5.171]$$

EXAMPLE 5.4.– For the BS model presented in section 5.6.3.2, the forward partial differential equation of Fokker–Planck [5.169] becomes here:

$$\frac{\partial p'}{\partial t}(x,y,t) = -\frac{\partial}{\partial y}\left[\delta p'(x,y,t\right] + \frac{1}{2}\frac{\partial}{\partial t}\left[\sigma^2(t,y)p'(x,y,t)\right]. \quad [5.172]$$

If

$$P[a < C(t) \leq b | C(t_0) = x_0] = \int_a^b p(x_0, t_0; y, t) dy, \qquad [5.173]$$

it can be shown that the process is time homogeneous, that is:

$$p(x_0, t, y, s) = p(x_0, x, s - t) \qquad [5.174]$$

The forward partial differential equation of Fokker–Planck [5.169] becomes here:

$$\frac{\partial p'}{\partial t}(x, y, t) = -\frac{\partial}{\partial y}[\delta p'(x, y, t] + \frac{1}{2}\frac{\partial}{\partial t}[\sigma^2(t, y) p'(x, y, t)]. \qquad [5.175]$$

It can be shown that:

$$p'(y, t; x_0, t_0) = \frac{1}{\sigma\sqrt{2\pi(t - t_0)}} e^{-\frac{\ln(x/x_0) - (\mu - \frac{1}{2}\sigma^2)(t - t_0)}{2\sigma^2(t - t_0)}}, \qquad [5.176]$$

result proving the lognormality distribution of $C(t)/C(t_0)$.

5.7. Multidimensional diffusion processes

We follow the notations of [JAN 13]: the state random vector ξ_t belongs to $\mathbb{R}^n$, $\mathbf{a}(x,t)$ belongs to $\mathbb{R}^n$ and $\mathbf{b}(x,t)$ belongs to an $m \times n$ real matrix, and all the components of $\mathbf{a}, \mathbf{b}$ are measurable functions.

Let us recall that we work with the matrix norm defined by:

$$M = (m_{ij}) \in R^{n \times m} : \|M\|^2 = \sum_{i=1}^{n} \sum_{j=1}^{m} m_{ij}^2. \qquad [5.177]$$

Let us consider a system whose state is given by the m-dimensional stochastic process $\xi(t) = (\xi_1(t), \ldots, \xi_n(t))'$ and is governed by the following stochastic differential system (SDS):

$$\begin{aligned} d\xi(t) &= \mathbf{a}(\xi, t) dt + \mathbf{b}(\xi, t) d\mathbf{B}(t), \\ \xi(0) &= \xi_0, \end{aligned} \qquad [5.178]$$

where $\mathbf{a}$ is an n-dimensional random vector of class L or D, $\mathbf{b}$ is a stochastic matrix nxm whose elements are measurable functions and $\mathbf{B}$ is an n-vector of n independent standard Brownian motions.

Propositions 5.11, 5.12 and 5.13 remain true provided that we work with the matrix norm defined by [5.176] (see [GIK 69]).

5.7.1. *Definition of multidimensional Itô and diffusion processes*

DEFINITION 5.11.– *The solution* $\xi = (\xi(t), t \in [0,T])$ *of [5.178] is called a multidimensional Itô process.*

DEFINITION 5.12.– A *multidimensional diffusion process* is defined as *m-dimensional Itô process* $\xi(t) = (\xi_1(t), \ldots, \xi_n(t))'$ *governed by the following SDS*:

$$\begin{aligned} d\xi(t) &= \mathbf{a}(\xi,t)dt + \mathbf{b}(\xi,t)d\mathbf{B}(t), \\ \xi(0) &= \xi_0, \end{aligned} \qquad [5.179]$$

where $\boldsymbol{a}$ *is an n-dimensional vector,* $\boldsymbol{b}$ *is a matrix nxm whose elements are measurable functions and continuous in the t variable and* $\boldsymbol{B}$ *is an n-vector of n independent standard Brownian motions satisfying the assumptions of extension of proposition 5.11. It follows that this diffusion process is a.s. unique.*

5.7.2. *Properties of multidimensional diffusion processes*

With matrix notations, we have the equivalent propositions of 5.13 and 5.14.

PROPOSITION 5.15 (MARKOV PROPERTY).– *Under the assumptions of the extension of proposition* 5.11 *and if, for each t,* $\Im_t$ *represents the* σ*-algebra generated by* ξ_0 *and the set* $(B(s), s \leq t))$*, then the diffusion process* $\xi = (\xi(t), 0 \leq t \leq T)$ *satisfies a.s.:*

$$P\left[\xi(t) \in A \middle| \Im_s\right] = P\left[\xi(t) \in A \middle| \xi(s)\right] (= p(s, \xi(s), t, A)) \qquad [5.180]$$

for all t > s and for all Borel set A.

PROPOSITION 5.16.– The function $p(s, \xi(s), t, A)$ of $\mathbb{R}^+ \times \mathbb{R}^n \times \mathbb{R}^+ \times \beta^{\otimes n} \mapsto [0,1]$ defined by [5.180] satisfies the following properties:

1) for all fixed $s, x, t, p(s, x, t, .)$ s is a probability measure on $\mathbb{R}^d$;

2) for all fixed s s,t,A, $p(s,.,t,A)$ is Borel-measurable;

3) the function p satisfies the Chapman–Kolmogorov equations:

$$\forall 0\leq s<t<\tau, x\in\mathbb{R}^d, A\in\beta^{\otimes d}:\\ \int_R p(s,x,t,dy)p(t,y,\tau,A)=p(s,x,\tau,A). \quad [5.181]$$

4) the process solution $\xi=(\xi(s),s\geq 0)$ is a Feller process, that is for all continuous bounded function of $\mathbb{R}\mapsto\mathbb{R}$, the application:

$$(s,x)\mapsto\int f(y)p(s,x,s+t,dy) \quad [5.182]$$

is continuous.

5) the process solution $\xi=(\xi(s),s\geq 0)$ satisfies the strong Markov property, that is [5.180] but where s and t are replaced by stopping times.

It can also be shown that:

$$1)\ \forall\varepsilon>0, t\geq 0, x\in\mathbb{R}^n: \lim_{h\downarrow 0}\frac{1}{h}\int_{|y-x|>\varepsilon} p(t,x,t+h,dy)=0, \quad [5.183]$$

$$2)\ \forall\varepsilon>0, t\geq 0, x\in\mathbb{R}^n:$$

$$\text{i)}\ \lim_{h\downarrow 0}\frac{1}{h}\int_{|y-x|<\varepsilon}(y-x)p(t,x,t+h,dy)=a(x,t),$$

$$\text{ii)}\ \lim_{h\downarrow 0}\frac{1}{h}\int_{|y-x|<\varepsilon}(y-x)(y-x)^{\tau}p(t,x,t+h,dy)=G(x,t), \quad [5.184]$$

$$G=bb^{T}.$$

As for the one-dimensional case, for the applications of such processes in finance, it is interesting to give the interpretations of these last properties:

1) The probability for the process $\xi=(\xi(s),s\geq 0)$ to have jumps of amplitude more than ε between t and $t+h$ is $o(h)$.

Consequently, the process $\xi=(\xi(s),s\geq 0)$ is continuous in probability.

2) Properties i) and ii) can be rewritten as follows:

$$\begin{aligned}&\text{i)}\quad E\left[\xi(t+h)-\xi(t)\middle|\xi(t)=\mathrm{x}\right]=\mathrm{a}(\mathrm{x},t)h+\mathrm{o}(h),\\&\text{ii)}\quad E\left[(\xi(t+h)-\xi(t))(\xi(t+h)-\xi(t))^{T}\middle|\xi(t)=\mathrm{x}\right]=(\mathrm{bb}^{T})(\mathrm{x},t)h+\mathrm{o}(h).\end{aligned}\qquad [5.185]$$

Consequently, the drift vector $\mathbf{a}$ *gives the rate of the conditional mean of the increment of the diffusion process on the infinitesimal time (t, t + h) interval and the matrix* $\mathbf{G}=\mathbf{bb}^{T}$, *the conditional variance–covariance matrix of this increment as the square of the mean vector is of order* $o(h)$.

5.7.3. Kolmogorov equations

Let:

$$p(s,\mathbf{x},t,\mathbf{y})=p\left(s,\mathbf{x},t,\left\{z:\xi(t)\leq\mathbf{y}\right\}\right)\qquad [5.186]$$

The probability transition distribution function $p(s,\mathbf{x},t,\mathbf{y})$ is a solution of *the prospective or forward partial differential equation of Fokker–Planck:*

$$\frac{\partial p}{\partial t}+\frac{\partial}{\partial x}(\mathrm{a}(\mathbf{x},t)p)-\frac{1}{2}\frac{\partial^2}{\partial x^2}(\mathbf{bb}^{\tau}(\mathbf{x},t)p)=0.\qquad [5.187]$$

with as initial condition:

$$\lim_{s\to t}p(x,t,y,s)=\begin{cases}1,x\leq y,\\0,x\neq y.\end{cases}\qquad [5.188]$$

If the function p has a density p', then it is a solution of *the prospective or forward partial differential equation of Fokker–Planck*:

$$\frac{\partial p'}{\partial t}+\frac{\partial}{\partial x}(\mathrm{a}(\mathrm{x},t)p')-\frac{1}{2}\frac{\partial^2}{\partial x^2}(\mathrm{bb}^{\tau}(\mathrm{x},t)p')=0.\qquad [5.189]$$

with as initial condition:

$$\lim_{s\to t}p'(\mathbf{x},t,\mathbf{y},s)=\begin{cases}1,\mathbf{x}=\mathbf{y},\\0,\mathbf{x}\neq\mathbf{y}.\end{cases}\qquad [5.190]$$

Here, the retrospective or backward equations take the following form:

$$\begin{aligned}-\frac{\partial p}{\partial t}&=\frac{1}{2}\mathbf{bb}^{\tau}(x,t)\frac{\partial^2 p}{\partial x^2}+\mathbf{a}(x,t)\frac{\partial p}{\partial x}\\-\frac{\partial p'}{\partial t}&=\frac{1}{2}\mathbf{bb}^{\tau}(x,t)\frac{\partial^2 p'}{\partial x^2}+\mathbf{a}(x,t)\frac{\partial p'}{\partial x}.\end{aligned} \quad [5.191]$$

with initial conditions, respectively:

$$\lim_{s\to t} p(\mathbf{x},t,\mathbf{y},s)=\begin{cases}1, \mathbf{x}\leq\mathbf{y},\\0, \mathbf{x}\neq\mathbf{y}.\end{cases} \quad [5.192]$$

$$\lim_{s\to t} p'(\mathbf{x},t,\mathbf{y},s)=\begin{cases}1, \mathbf{x}=\mathbf{y},\\0, \mathbf{x}\neq\mathbf{y}.\end{cases}$$

To describe these two partial derivative equations (PDEs), let us introduce the following two functional operators:

$$\begin{aligned}L&=\sum_i a_i(s,\mathbf{x})\frac{\partial}{\partial x_i}+\frac{1}{2}\sum_{i,j} b_{ij}(s,\mathbf{x})\frac{\partial^2}{\partial x_i\partial x_j},\\L^*&=\sum_i\frac{\partial(a_i(s,\mathbf{x}),.)}{\partial y_i}-\frac{1}{2}\sum_{i,j} b_{ij}(s,x)\frac{\partial^2(b_{ij}(s,\mathbf{x}),.)}{\partial y_i\partial y_j},\end{aligned} \quad [5.193]$$

with L^* being the adjoint operator of L.

Now, let g be a bounded measurable scalar function and define the function u as follows:

$$u(s,\mathbf{x})=E\left[g(\xi(t)\middle|\xi(s)=\mathbf{x}\right] \quad [5.194]$$

and so:

$$u(s,\mathbf{x})=\int_{R^n} g(\mathbf{y})P(s,\mathbf{x},t,\mathbf{dy}). \quad [5.195]$$

It can be shown (see [GAR 88]) that the function u is differentiable with respect to s and that u has partial derivative of the first and second order with respect to the $x_i, i=1,...,n$ and moreover satisfies the PDE:

$$\begin{aligned}&\frac{\partial u}{\partial s}+Lu=0,\\&\lim_{s\nearrow t} u(s,\mathbf{x})=g(\mathbf{x}),\end{aligned} \quad [5.196]$$

is called the Kolmogorov's *backward equation*, also called *backward* because it concerns the "backward" variables s and $\boldsymbol{a}$.

If p has a density p', the PDE of [5.189] with an initial condition:

$$\lim_{s \nearrow t} p'(s,\mathbf{x},t,\mathbf{y}) = \delta(\mathbf{x}-\mathbf{y}) \quad [5.197]$$

is equivalent to:

$$\begin{aligned} &\frac{\partial p'}{\partial s} + Lp = 0, \\ &\lim_{s \nearrow t} p'(s,\mathbf{x},t,\mathbf{y}) = \delta(\mathbf{x}-\mathbf{y}), \end{aligned} \quad [5.198]$$

and similarly with the operator L^* for the forward Kolmogorov equation with:

$$\lim_{s \to t} p'(\mathbf{x},t,\mathbf{y},s) = \begin{cases} 1, \mathbf{x} = \mathbf{y}, \\ 0, \mathbf{x} \neq \mathbf{y}. \end{cases} \quad [5.199]$$

EXAMPLE 5.5.– The standard Brownian motion satisfies the SDE [5.179] with **a** as vector null and matrix **B** as the identity matrix. Consequently, PDE [5.189] *becomes for* the backward equation, we get:

$$\frac{\partial p'}{\partial s} + \frac{1}{2}\sum_{i=1}^{n} \frac{\partial^2 p'}{\partial^2 x_i^2} = 0 \quad [5.200]$$

and for the forward case, we have:

$$\frac{\partial p'}{\partial t} - \frac{1}{2}\sum_{i=1}^{n} \frac{\partial^2 p'}{\partial^2 y_i^2} = 0, \quad [5.201]$$

with [5.199] as an initial condition.

It can be shown that the solution is given by:

$$p'(,\mathbf{x},s,t,\mathbf{y}) = \frac{1}{(2\pi(t-s))^{\frac{n}{2}}} e^{-\frac{\|\mathbf{y}-\mathbf{x}\|^2}{2(t-s)}}. \quad [5.202]$$

5.7.4. *The Stroock–Varadhan martingale characterization of diffusion processes [KAR 81]*

Let us consider ξ be a time homogeneous diffusion process characterized by the following SDE:

$$\begin{aligned} d\xi(t) &= \mu(\xi(t))dt + \sigma(\xi(t))dB(t), \\ \xi(0) &= \xi_0, \end{aligned} \qquad [5.203]$$

and define, for every $\lambda > 0$, the following stochastic process X:

$$X(t) = \exp\left\{\lambda\xi(t) - \lambda\int_0^t \mu(\xi(s))ds - \frac{1}{2}\lambda^2\int_0^t \sigma^2(\xi(s))ds\right\}, t > 0. \qquad [5.204]$$

The main result of Stroock–Varadhan is that, *under regular assumptions, the process X is a martingale with respect to the filtration generated by the Brownian motion B and conversely: if, for every λ, X is a martingale with respect to the filtration generated by the Brownian motion B, then the process ξ is a diffusion process.*

In particular, if the process ξ is a standard Brownian motion, result [5.204] shows that the process defined by:

$$Y(t) = \exp\left\{\lambda B(t) - \frac{\lambda t^2}{2}\right\} \qquad [5.205]$$

is a martingale.

Another characterization of a homogeneous diffusion process is given as follows: the process ξ is a homogeneous diffusion process of parameters $\mu(x), \sigma(x)$ *iff* for every function *f*, bounded and twice continuous differentiable, the process defined by:

$$Z(t) = f(\xi(t)) - f(\xi(0)) - \int_0^t \left[\frac{1}{2}\sigma^2(\xi(s))f''(\xi(s)) + \mu(\xi(s))f'(\xi(s))\right]ds \qquad [5.206]$$

is a martingale.

For example, for ξ being a standard Brownian motion, with $f(x) \equiv x, f(x) \equiv x^2$, we, respectively, find that the processes ξ and $(\xi^2(t) - t, t \in [0,T]$ are martingales, i.e. results of properties 5.2 and 5.3.

5.8. Relation between the resolution of PDE and SDE problems. The Feynman–Kac formula [PLA 06]

This relation is essential as indeed, it shows the strong interaction between the resolution of PDE and SDE. Solving on problem gives the solution of the corresponding other one and reciprocally. Consequently, we will have two ways of solving the problems in these two fields both theoretically and numerically. The same is also true for simulation. We will present the fundamental result of *Feynman–Kac.*

5.8.1. *Terminal payoff*

On the filtered probability space $\left(\Omega,\Im,\left(\Im_t,t\geq 0\right),P\right)$, let us consider an n-dimensional diffusion process $\xi_{\mathbf{x}}$ on $\left[t,T\right],t<T$ such that:

$$\xi(t)=\mathbf{x} \tag{5.207}$$

Furthermore, we give a function $H:\mathbb{R}^n\mapsto\left[0,\infty\right)$ such that:

$$E\left(\left|H(\xi_{\mathbf{x}}(T)\right|\right)<\infty. \tag{5.208}$$

Now, let us consider another function $u:\left[0,T\right]\times\mathbb{R}^n\mapsto\left[0,\infty\right)$ satisfying on $(0,T)\times R^n$ the following PDE:

$$\frac{\partial u}{\partial t}(\mathbf{x},t)+\sum_{i=1}^{n}a_i(x,t)\frac{\partial u}{\partial x_i}(\mathbf{x},t)+\frac{1}{2}\sum_{i,k=1}^{n}\sum_{j=1}^{m}b_{ij}(\mathbf{x},t)b_{kj}(\mathbf{x},t)\frac{\partial^2}{\partial x_i\partial x_k}u(\mathbf{x},t)=0$$

or [5.209]

$$L^*(u(x,t))=0,L^*=L+\frac{\partial}{\partial t}$$

with the Cauchy problem defined by the following *terminal condition*:

$$u(\mathbf{x},T)=H(\mathbf{x}),\mathbf{x}\in\mathbb{R}^n. \tag{5.210}$$

The *Feynman–Kac formula* asserts that the solution u takes the form:

$$\begin{gathered}u(\mathbf{x},t)=E\left(H(\xi_{\mathbf{x}}(T)\middle|\Im_t\right),\\ \forall\mathbf{x}\in\mathbb{R}^n,\forall t\in\left[0,T\right].\end{gathered} \tag{5.211}$$

5.8.2. *Discounted payoff function*

For the following, it is necessary to extend [5.200] as follows:

$$\frac{\partial u}{\partial t}(\mathbf{x},t)+\sum_{i=1}^{n} a_i(x,t)\frac{\partial u}{\partial x_i}(\mathbf{x},t)+\frac{1}{2}\sum_{i,k=1}^{n}\sum_{j=1}^{m} b_{ij}(\mathbf{x},t)b_{kj}(\mathbf{x},t)\frac{\partial^2}{\partial x_i \partial x_k}u(\mathbf{x},t)$$
$$= r(\mathbf{x},t)u(\mathbf{x},t) \qquad [5.212]$$

or

$$L^*(u(\mathbf{x},t)) = r(\mathbf{x},t)u(\mathbf{x},t)$$

where the function $r:\mathbb{R}^n \times [0,T]$ will represent later the short-term interest rate function.

The terminal condition is still given by relation [5.210]. Thus, under regular assumptions, the unique solution of this Cauchy problem is given by *the following so-called Feynman–Kac formula*:

$$u(\mathbf{x},t) = E\left(\exp(e^{-\int_t^T r(\xi_{\mathbf{x}}(s),s)ds})H(\xi_{\mathbf{x}}(T)\middle|\Im_t\right), \qquad [5.213]$$
$$\forall \mathbf{x}\in\mathbb{R}^n, \forall t\in[0,T].$$

5.8.3. *Discounted payoff function and payoff rate*

To the last problem, we add another function $g:\mathbb{R}^n \times [0,T]$, representing in finance a continuous payment on $[0,T]$ at a rate of g per unit time.

The extension of PDE [5.209] is now given by:

$$\frac{\partial u}{\partial t}(\mathbf{x},t)+\sum_{i=1}^{n} a_i(\mathbf{x},t)\frac{\partial u}{\partial x_i}(\mathbf{x},t)+\frac{1}{2}\sum_{i,k=1}^{n}\sum_{j=1}^{m} b_{ij}(\mathbf{x},t)b_{kj}(\mathbf{x},t)\frac{\partial^2}{\partial x_i \partial x_k}u(\mathbf{x},t)$$
$$+g(\mathbf{x},t) = r(\mathbf{x},t)u(\mathbf{x},t) \qquad [5.214]$$

or

$$L^*(u(\mathbf{x},t)) + g(\mathbf{x},t) = r(\mathbf{x},t)u(\mathbf{x},t)$$

always with the terminal condition [5.210].

Then under regular assumptions, which we will discuss later on, the unique solution of this Cauchy problem is given by *the following so-called Feynman–Kac formula:*

$$u(\mathbf{x},t) = E\left(H(\xi_{\mathbf{x}}(T)\exp\left(-\int_t^T r(\xi_{\mathbf{x}}(s),s)ds\right) + \int_t^T g(\xi_{\mathbf{x}}(s),s)ds \exp\left(-\int_t^s r(\xi_{\mathbf{x}}(z),z)dz)\right) \| \Im_t \right), \quad [5.215]$$

$\forall \mathbf{x} \in \mathbb{R}^n, \forall t \in [0,T].$

REMARK 5.10.– It must be clear that the dependence on $\mathbf{x}$ in the Feynman–Kac formulas [5.210], [5.212] and [5.214] came from the initial condition on the system at time t given by relation [5.207].

Platen and Heath [PLA 06, pp. 362–363], give sufficient conditions so that the preceding Feynman–Kac formulas hold. Another presentation of such conditions is given in [MAO 97, p. 82].

EXAMPLE 5.6.– The heat equation

This equation has the form:

$$\frac{\partial u}{\partial t} = \Delta u,$$

where [5.216]

$$\Delta u(\mathbf{x},t) = (\sum_{i=1}^n \frac{\partial^2}{\partial x_i^2})u(\mathbf{x},t).$$

The initial condition defining the Cauchy problem is given by:

$$u(\mathbf{x},0) = f(\mathbf{x}), \mathbf{x} \in \mathbb{R}^n. \quad [5.217]$$

For this problem, Itô's process defined by relation [5.203] here becomes:

$$d\xi_x(t) = d\mathbf{B}(t), \qquad [5.218]$$
$$\xi_x(0) = \mathbf{x}.$$

It follows that:

$$\xi_x(t) = \mathbf{x} + \mathbf{B}(t)$$

As here we have instead of $[t,T],[0,t]$, from result [5.211], we get:

$$u(\mathbf{x},t) = E_x(f(\mathbf{x} + \mathbf{B}(t)), \tag{5.219}$$

where $E_{\mathbf{x}}$ represents the expectation relative to the probability measure generated by the standard Brownian motion starting from $\mathbf{x}$ at time t.

For $n = 1$, we get:

$$\begin{aligned} &\frac{\partial u}{\partial t}(x,t) = \frac{\partial^2 u}{\partial x^2}(x,t) \\ &u(x,t) = E_x(f(x + B(t)), \\ &\qquad = \frac{1}{\sqrt{2\pi t}} \int_{-\infty}^{+\infty} f(x+s) e^{-\frac{s^2}{2t}} ds. \end{aligned} \tag{5.220}$$

With the change of variable $x + s \mapsto y$, we finally get:

$$u(x,t) = \frac{1}{\sqrt{2\pi t}} \int_{-\infty}^{+\infty} f(y) e^{-\frac{(y-x)^2}{4t}} dy \tag{5.221}$$

EXAMPLE 5.7.– The Feynman–Kac equation

This PDE is given by:

$$\frac{\partial u}{\partial t} = \Delta u + cu, \tag{5.222}$$

where c is a function only of x. Here, we have a solution of the Cauchy problem defined still by relation [5.218].

EXAMPLE 5.8.– The Cameron–Martin equation

This PDE is given by:

$$\frac{\partial u}{\partial t} = \Delta u + \langle \mathbf{c}, \mathbf{grad} u \rangle, \tag{5.223}$$

where **a** is only a function of **x**:

$$\mathbf{a}(\mathbf{x}) = (a_1(\mathbf{x}), \ldots, a_n(\mathbf{x})). \quad [5.224]$$

Here, the solution of the Cauchy problem defined still by [5.217] as initial condition is given by:

$$u(\mathbf{x},t) = E_x\left(f(\mathbf{x}+\mathbf{B}(t)) e^{\int_0^t \mathbf{a}(\mathbf{x}+\mathbf{B}(s))ds - \frac{1}{2}\int_0^t \|\mathbf{a}(s)\|^2 ds} \right). \quad [5.225]$$

5.9. Application to option theory

5.9.1. *Options*

During the last 30 years, financial innovation has generalized the systematic use of new financial instruments such as *options* and *swaps*, mainly motivated for hedging but also, sometimes, used as speculative tools.

So, let us give some basic concepts of *option theory*.

DEFINITION 5.13.– *A call option (respectively, put option) is a contract giving the right to buy (respectively, to sell) a financial asset, called an underlying asset, for a fixed price, called exercise price, at the end of the contract time, called maturity time, also laid down in the contract.*

If we can exercise the option at any time before maturity, this type of option is called an *American type*; if we can exercise it only at maturity, the option is called a *European type*.

Let us use the following notation:

– K: exercise price;

– T: maturity time;

– S: value of the underlying asset at maturity.

Thus, the "gain" of the holder of a European option at maturity time T is represented by the following graph.

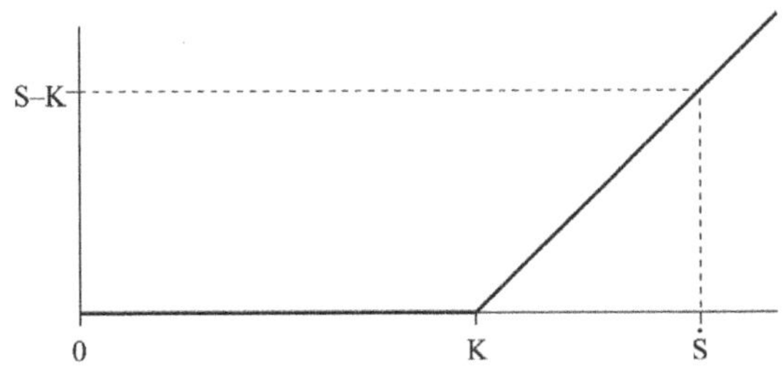

Figure 5.3. *Holder's gain at maturity for call option*

For the holder of a put, this graph becomes:

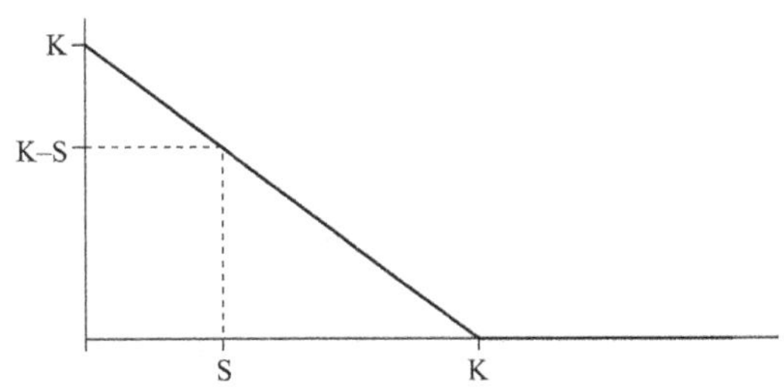

Figure 5.4. *Holder's gain at maturity for put option*

Of course, to get the "net gain", we must estimate the cost of the option, often called *option premium*, and furthermore transaction costs and taxes.

Let us represent, respectively, by C and P the premiums of call and put options.

So, we get, without taking into account transaction costs and taxes, the following two graphical representations

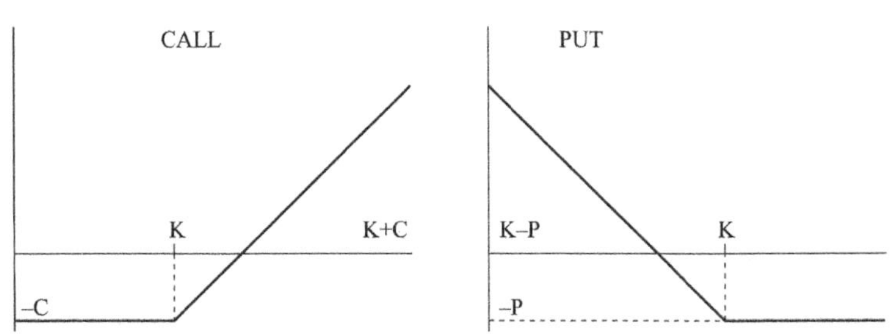

Figure 5.5. *Net gains at maturity for the holder of call and put options*

The main problem is called the *pricing of optional products,* that is to give within the framework of an economic-financial theory, the values of premiums C and P as a function of the maturity T and the value of the asset at time 0.

More generally, as the holder of an option can sell his option on the option market at any time t, $0<t<T$, it is also necessary to give the "fair" value of the option at this time t knowing that the underlying asset has, at this time, the value $S = S(t)$.

Commonly, this fair market value is represented by:

$$C(S,\tau) \tag{5.226}$$

where $\tau = T - t$ represents the maturity computed at time t.

Sometimes, it is also useful to represent the call value as a function of the time $C(S,t)$.

We see here that it is absolutely necessary to make assumptions about the stochastic process $S = \left(S(t), 0 \leq t \leq T\right)$.

Concerning the economic-financial theory framework, we adopt the assumption of *efficient market,* meaning that all the information available at time t is reflected in the values of the assets and so, transactions having an abnormally high profitability are not possible.

More precisely, an *efficient market* satisfies the following assumptions:

1) absence of transaction costs;

2) possibility of short sales;

3) availability of all information to all the economic agents;

4) perfect divisibility of assets;

6) continuous time financial market.

Furthermore, the market is *complete*; meaning that there exist zero-coupon bonds without risk for all possible maturities.

5.9.2. *Black and Scholes model*

In fact, Black and Scholes used a continuous time model for the underlying asset process S introduced by Samuelson [SAM 65] that is nothing else but a very simple diffusion process given by:

$$\begin{aligned} &dS(t) = \mu S(t)dt + \sigma S(t)dB(t), t \in [0,T] \\ &S(0) = S_0, \end{aligned} \tag{5.227}$$

This is the Black–Scholes–Samuelson process introduced in section 5.6.2.2. With the notations of relation [5.120], we have:

$$\xi(t) = S(t), \mu(\xi,t) = \mu, \sigma(\xi,t) = \sigma.$$

Concerning the economic-financial theory framework, they adopt the assumption of *efficient market,* meaning that all the information available at time t is reflected in the values of the assets and therefore, transactions having an abnormally high profitability are not possible.

5.9.3. *The Black and Scholes partial differential equation (BSPDE) and the BS formula*

Under the above assumptions and using a principle of absence of arbitrage possibility, Black and Scholes obtained the following PDE satisfied by C(S,t):

$$-rC(S,t) + r\frac{\partial C}{\partial S}(S,t)S + \frac{\partial C}{\partial t}(S,t) + \frac{1}{2}\frac{\partial^2 C}{\partial^2 S^2}(S,t)\sigma^2 S^2 = 0, \tag{5.228}$$

with as initial condition

$$C(S,t) = \begin{cases} 0, t \in [0,T), \\ \max\{0, S-K\}, t = T. \end{cases}$$

and where r is the non-risky interest rate.

Comparing with the particular case of PDE [5.211] with n=1, that is:

$$\frac{\partial u}{\partial t}(x,t) + a(x,t)\frac{\partial u}{\partial x}(x,t) + \frac{1}{2}b^2(x,t)\frac{\partial^2 u}{\partial x^2} = r(x,t)u(x,t) \tag{5.229}$$

The identification gives:

$$x \equiv S, r(S,t) \equiv r, a(S,t) \equiv rS, b(S,t) \equiv \sigma^2 S^2. \qquad [5.230]$$

Consequently, we can use result [5.212] to express the solution under the following form:

$$u(x,t) = E\left(\exp(e^{-\int_t^T r(\xi(s),s)ds}) H(\xi_{\mathbf{x}}(T) | \Im_t \right),$$

$$\forall \mathbf{x} \in \mathbb{R}^n, \forall t \in [0,T], \qquad [5.231]$$

where here: $x = S, u(x,t) = C(S,t), \xi(s) = S(s), \xi_x(T) = S_{S_0}(T) = S(T),$

$$H(\xi_{\mathbf{x}}(T) = \max\{0, S(T) - K\}.$$

Finally, we obtain the solution under the form:

$$u(x,t) = E\left(e^{-\int_t^T rds} \max\{0, \xi_{\mathbf{x}}(T) - K\} | \Im_t \right), \qquad [5.232]$$

As the instantaneous rate r is constant, we finally obtain returning to the option theory notations:

$$C(S,t) = e^{-r(T-t} E\left(\max\{0, S(T) - K\} | \Im_t\right). \qquad [5.233]$$

Now, the expectation in [5.233] must be taken in the probability measure on which is defined the diffusion process:

$$d\xi(t) = a(\xi,t)dt + \mathrm{b}(\xi,t)dB(t),$$
$$\xi(0) = \xi_0, \qquad [5.234]$$

or with [5.230]

$$d\xi(t) = r\xi dt + \sigma\xi dB(t),$$
$$\xi(0) = \xi_0. \qquad [5.235]$$

This last SDE is different from [5.227] as the trend μ has been replaced by the instantaneous non-risky rate r and thus the diffusion process is now defined on a new probability space $(\Omega,\Im,(\Im_t),Q)$. Below, we will show how pass from the initial measure P to the new measure Q. Nevertheless, from result [5.161], we know that:

$$\xi(t)=\xi_0 e^{(r-\frac{\sigma^2}{2})t+\sigma B(t)}$$

and so [5.236]

$$\ln\frac{\xi(t)}{\xi_0}=(r-\frac{\sigma^2}{2})t+\sigma B(t).$$

and thus it follows that we know the distribution of $\xi(T)$ giving the possibility to compute the expectation [5.232]. Coming back to the option theory notation, this computation gives the following result:

$$C(S,t)=S\Phi(d_1)-Ke^{-r(T-t)}\Phi(d_2)$$

where

$$d_1=\frac{1}{\sigma\sqrt{T-t}}\left[\ln(\frac{S}{K})+(r+\frac{\sigma^2}{2})(T-t)\right],$$ [5.237]

$$d_2=d_1-\sigma\sqrt{T-t}.$$

The result is known as the *Black and Scholes formula.*

REMARK 5.11.– For the put option, we obtain the following result:

$$P(S,t)=Ke^{-r(T-t)}\Phi(-d_2)-S\Phi(-d_1),$$

$$d_1=\frac{1}{\sigma\sqrt{T-t}}\left[\log\frac{S}{K}+\left(r+\frac{\sigma^2}{2}\right)(T-t)\right],$$

$$d_2=d_1-\sigma\sqrt{T-t},$$

$$S=S(t).$$

as a result of call–put parity relation (see, for example, [JAN 09b]):

$$C(S,t)-P(S,t)=S-e^{-r(T-t)}K.$$

EXAMPLE 5.9.– Table 5.1 gives the data and results for pricing call and put options.

Market data		
Weekly volatility		0,02773501
volatility		0,25
T		1
Asset price at t=0		100
Exercise price		100
instantaneous interest rate		0,009950331
Yearly interest rate		0,01
Black & Scholes results		
d1		0,164801323
d2		-0,085198677
y		0,963228843
z		1,020132918
N(d1)		0,393561367
N(d2)		0,397496981
Phi(d1)		0,565449826
Phi(d2)		0,466051201
CALL price		10,4012993
PUT price		9,41120029

Table 5.1. *Example of application of the BS formula*

5.9.4. *Girsanov theorem*

5.9.4.1. *Radon-Nikodym derivative*

On the filtered complete space $(\Omega, \Im, (\Im_t))$, let us consider two probabilities measures P and Q.

The measure Q is *absolutely continuous* with respect to the measure P if for every set of $\Im$ such that $P(A)=0$, then $Q(A)=0$.

In this case, there exists a r.v. Z denoted by $Z = \frac{dQ}{dP}$ and is called the *Radon–Nikodym derivative of Q with respect to P* such that for all integrable r.v. X, and for all A belonging to $\Im$ we have:

$$\int_A XdQ = \int_A X\frac{dQ}{dP}dP (= \int_A XZdP.$$

5.9.4.2. *The Girsanov theorem*

In its simplified form (see [GIK 75]), this theorem affirms that:

On the filtered probability space $(\Omega, \Im, (\Im_t), P)$, *let B be an SBM and* $f = \{f(t), 0 \le t \le T\}$ *be an adapted stochastic process such that:*

$$E_P[e^{\frac{1}{2}\int_0^T f^2(s)ds}] < \infty \text{. (Novikov's condition)} \qquad [5.238]$$

Then, there exists a unique new probability measure equivalent to the initial measure P on $(\Omega, \Im, (\Im_t), P)$ *with Radon–Nikodym density as:*

$$\frac{dQ}{dP} = \exp\left(-\int_0^T f(s)dB(s) - \frac{1}{2}\int_0^T f^2(s)ds\right), \qquad [5.239]$$

such that, under this new measure Q, the stochastic process $\hat{B}$ *defined by:*

$$\hat{B} = \left\{\hat{B}(t), 0 \le t \le T\right\} \qquad [5.240]$$

and

$$\hat{B}(t) = B(t) - \int_0^t f(s)ds \qquad [5.241]$$

is standard Brownian motion on [0,T] on the probability space $(\Omega, \Im, (\Im_t), Q)$.

Moreover, we have:

$$\frac{dQ(f)}{dP} = \xi(T),$$

$$\int X(\omega)dQ(f) = \int X(\omega)\xi(T,\omega)dP$$

or [5.242]

$$E_Q[X(\omega))] = E_P[X(\omega)\xi(T,\omega)],$$

$$E_P[X(\omega)] = E_Q\left[X(\omega)(\xi(T,\omega))^{-1}\right]$$

with

$$\zeta(T) = \exp\left(-\int_0^T f(s)dB(s) - \frac{1}{2}\int_0^T f^2(s)ds\right). \qquad [5.243]$$

REMARK 5.12.– As $Q(f)$ is a probability measure, if we integrate $dQ(f)$ on Ω the second equality of relation [5.242] shows that necessarily we must have:

$$E_P[\zeta(T)] = E_P\left[\exp\left(-\int_0^T f(s)dB(s) - \frac{1}{2}\int_0^T f^2(s)ds\right)\right] = 1. \qquad [5.244]$$

In fact, this last condition suffices to prove Girsanov's theorem but Novikov's condition is, at least in finance, easier to verify.

5.9.5. *The risk-neutral measure and the martingale property*

It is now possible to construct another probability measure Q on $(\Omega, \Im, (\Im_t))$, called the *risk-neutral measure*, such that the value of the call given by formula [5.237] is simply the expectation value of the present value of the "gain" at maturity time T which is already proved by the Feynman–Fac formula.

To define this new measure, let us introduce the following constant function:

$$f(s) = -\frac{\mu - r}{\sigma}(= -\theta), \qquad [5.245]$$

θ representing, from the financial point of view, the *normalized risk premium*. It is clear that this function satisfies Novikov's condition [5.238] and thus the theorem of Girsanov can be applied. Relation [5.243] becomes:

$$\hat{B}(t) = B(t) + (\frac{\mu - r}{\sigma})t,$$

$$d\tilde{B}(t) = dB(t) + (\frac{\mu - r}{\sigma})dt. \qquad [5.246]$$

The BSB SDE [5.229] takes the following form:

$$dS = \mu S dt + \sigma S[d\tilde{B}(t) - (\frac{\mu - r}{\sigma})dt],$$

or [5.247]

$$dS = rSdt + \sigma S d\tilde{B}(t).$$

This is the result [5.235] given by the application of the Feynman–Kac formula.

Here, the stochastic process S must be considered as a stochastic process $(S(t), t \in [0,T])$ on the probability space $(\Omega, \Im, (\Im_t), Q)$, the Radon–Nikodym density of Q being given by relation [5.242]:

$$\begin{aligned}\frac{dQ}{dP} &= \exp\left(-\int_0^T \frac{\mu - r}{\sigma} dB(s) - \frac{1}{2}\int_0^T (\frac{\mu - r}{\sigma})^2 ds\right),\\ &= \exp\left(-\frac{\mu - r}{\sigma} B(T) - \frac{1}{2}(\frac{\mu - r}{\sigma})^2 T\right).\end{aligned} \quad [5.248]$$

Let us now introduce the present values computed with rate r at time 0 of the asset values defining a new stochastic process $(\tilde{S}(t), t \in [0,T])$ where:

$$\tilde{S}(t) = e^{-rt} S(t). \quad [5.249]$$

On $(\Omega, \Im, (\Im_t), Q)$, we obtain:

$$\begin{aligned}d\tilde{S} &= -re^{-rt} S dt + e^{-rt} dS,\\ &= -r\tilde{S}dt + e^{-rt} dS,\\ &= -r\tilde{S}dt + e^{-rt}(rSdt + \sigma S d\tilde{B}(t),\\ &= -r\tilde{S}dt + r\tilde{S}dt + \sigma\tilde{S}d\tilde{B}(t),\\ &= \sigma\tilde{S}d\tilde{B}(t).\end{aligned}$$

As $\tilde{S}(0) = S(0) = S_0$, we have:

$$\tilde{S}(t) = S_0 e^{-\frac{\sigma^2}{2}t + \sigma\tilde{B}(t)}. \quad [5.250]$$

Let us now compute $E[\tilde{S}(t) | \tilde{S}(u), u \le s], (s < t)$.

Using relation [5.250] and the property of independent increments of $\tilde{B}$, we get:

$$\begin{aligned} E[\tilde{S}(t)|\tilde{S}(u), u \leq s] &= S_0 E[e^{-\frac{\sigma^2}{2}t+\sigma\tilde{B}(t)}|\tilde{B}(s)], \\ &= S_0 E[e^{-\frac{\sigma^2}{2}t+\sigma\tilde{B}(s)+\sigma(\tilde{B}(t)-\tilde{B}(s))}]. \end{aligned} \quad [5.251]$$

As $\left(\tilde{B}(t)-\tilde{B}(s)\right)$ is $N(0,\sigma(t-s))$, we know that:

$$E\left[e^{\sigma\left(\tilde{B}(t)-\tilde{B}(s)\right)}\right] = e^{\frac{\sigma^2(t-s)}{2}}.$$

And therefore, returning to relation [5.251], we finally have the martingale property, that is:

$$\begin{aligned} E[\tilde{S}(t)|\tilde{S}(u), u \leq s] &= S_0^{-\frac{\sigma^2}{2}s+\sigma\tilde{B}(s)}, \\ &= \tilde{S}(s). \end{aligned} \quad [5.252]$$

The exponential form of $\tilde{S}$ is the reason why we speak about *exponential martingale* on $(\Omega, \Im, (\Im_t), Q)$.

REMARK 5.13.– It is important to point out that the *risk-neutral measure* Q is an artificial measure introduced to facilitate the evaluation of derivative products as we will discuss later on, but in any case, it corresponds to the probability measure having a physical sense. Indeed, to evaluate some probabilities of events related to the process S, only the measure P must be used; this is why P is called the *historical* or *physical (probability) measure*.

Let us also remark, that under P, processes S or $\tilde{S}$ are not martingales as, for example, we have:

$$\begin{aligned} E\left[S(T)\right] &= S(0)e^{\mu T}, \\ E\left[\tilde{S}(T)\right] &= S(0)e^{(\mu-r)T}. \end{aligned} \quad [5.253]$$

In the last equality, the term $(\mu - r)$ is called the *risk premium* in basic finance. In a healthy economy, in general, this premium is positive, showing the interest to invest in shares instead of non-risky assets, but if the economic situation is not good, it is clearly possible that this premium becomes negative and in this case, investors will prefer non-risky assets provided that the *default risk* is low.

Let us say that the default risk for a society is the risk that it will not be possible for it to reimburse the interest amounts to the investors; it is measured by grades given by the notation agencies and the recent EURO crisis has well shown that the way they are working and acting must be seriously improved.

5.9.6. *The risk-neutral measure and the evaluation of derivative products*

From relations [5.251] and [5.252], we know that:

$$S(t) = S_0 e^{\left(r - \frac{\sigma^2}{2}\right)t} e^{\sigma \tilde{B}(t)}. \qquad [5.254]$$

Let us compute, under Q, the expectation of the present value of the "gain" an investor expects with a European call, with time horizon T, that is:

$$e^{-rT} E_Q \left(\sup\{S(T) - K, 0\}\right). \qquad [5.255]$$

So we have:

$$e^{-rT} E_Q \left(\sup\{S(T) - K, 0\}\right) = e^{-rT} E_Q \left[\sup\left\{S_0 e^{rT - \frac{1}{2}\sigma^2 T + \sigma B(T)} r^2 - K, 0\right\}\right].$$

result giving well the BS formula again (see [JAN 13], Chapter 5).

To conclude, we see that the knowledge of the risk-neutral measure – equivalent from the point of view of computation to the Feynman–Kac formula application – can replace the resolution of the partial derivative equation and replaces it by the calculation of an expectation, which is in general easier, as it only uses the marginal distribution of $S(T)$.

However, we must add that, for more complicated derivative products, it may be more interesting, from the numerical point of view, to solve this partial derivative equation with the finite difference method, and particularly in the case of American options.

More generally, under the risk-neutral measure, for every derivative product of the European type with gain and T given by:

$$C(S,T) = \psi(S,T), \quad [5.256]$$

where ψ is a given function, its price at T=0 can be computed with the following extension of formula [5.255]

$$C(S,0) = e^{-rT} E_Q\left(\psi(T)\right). \quad [5.257]$$

6

Lévy Processes

Lévy processes were introduced in 1962 by B. Mandelbrot as an alternative to Gaussian models. Today, some academics [SCH 03, EBE 99] consider these processes as a reply to the principal critiques of the Black and Scholes model essentially based on the lognormality of the stock values, which is of course untrue, in particular, in a period of crisis and intense speculation.

6.1. Notion of characteristic functions

For a random variable (r.v.) X of distribution function (d.f.) F, let us recall that the *characteristic function* (c.f.) (see [LUK 70]) of X is given by the following function ϕ:

$$\phi(t) = E\left[e^{itX}\right] = \int_{-\infty}^{+\infty} e^{iux} dF(x) (= \int_{-\infty}^{+\infty} e^{iux} f(x) dx \text{ if } f(x) = F'(x)), u \in \mathbb{R} \quad [6.1]$$

with $i = \sqrt{-1}$.

When it exists, the *generating function* φ of X is defined by:

$$\varphi(u) = \phi(-iu) = \int_{-\infty}^{+\infty} e^{ux} dF(x) (= \int_{-\infty}^{+\infty} e^{ux} f(x) dx \; si \; f(x) = F'(x)), u \in \mathbb{R},$$

Contrary to the generating function, the c.f. of X always exists for any value of u. Furthermore, there is a correspondence one-to-one between distribution function and characteristics of a random variable X.

If the two r.v. X and Y are independent having ϕ_X, ϕ_Y as c.f., the sum $X + Y$ has $\phi_X \times \phi_Y$ as the c.f.

Table 6.1 shows the c.f. of some d.f.

Distribution	Characteristic function
Poisson λ	$e^{\lambda(e^{iu}-1)}$
Normal (μ,σ^2)	$e^{i\mu u-\frac{\sigma^2u^2}{2}}$
Exponential λ	$\dfrac{1}{1-\dfrac{iu}{\lambda}}$
Gamma (a,b)	$\dfrac{1}{(1-\dfrac{iu}{b})^a}$

Table 6.1. *Characteristic functions of some distributions*

6.2. Lévy processes

As mentioned in the Introduction, the main fact of Lévy processes is to drop the normality assumption in the definition of a standard Brownian motion (SBM) and keep its other assumptions, with another slight assumption replacing the normality.

The basic space is a complete filtered probability space $(\Omega,\Im,(\Im_t,t\geq 0)),P)$.

DEFINITION 6.1.– *The stochastic process* $L=(L_t,t\in[0,\infty])$ *is called a Lévy process if it satisfies the following conditions*:

1) L is adapted to the basic filtration;

2) L has independent increments, i.e.:

$$\forall s,t\ (0\leq s<t)\ :\ \mathrm{P}\left[L_t-L_s\in A|\Im_s\right]=P\left[L_t-L_s\in A\right],\quad \forall \mathrm{A}\in\Im, \tag{6.2}$$

3) L has stationary increments, that is:

$$\forall s,t\ (0\leq s<t):\ \mathrm{L}_t-L_s\prec L_{t-u}-L_{s-u},\forall 0<u\leq s, \tag{6.3}$$

4) $L(0)=0$ a.s.;

5) X is stochastically continuous for every t, that is:

$$\forall\varepsilon>0:\lim_{\Delta t\to 0}P(|L_{t+\Delta t}-L_t|>\varepsilon)=0. \tag{6.4}$$

The concept of Lévy process is strongly related to that of a random variable X *infinitely divisible distribution* asserting that, for all $n \in \mathbb{N}_0$, there exist n independent and identically distributed (i.i.d.) random variables $Y_1^{(n)},...,Y_n^{(n)}$ such that $X = Y_1^{(n)} + ... + Y_n^{(n)}$.

This is equivalent to the following property:

$$\phi_x(t) = \phi_{Y_1^{(n)}}(t)...\phi_{Y_n^{(n)}}(t) \qquad [6.5]$$

Of course, this is the case of the normal probability law because if $X_1,...,X_n$ are independent r.v. $N(m_1,\sigma_1^2),...,N(m_n,\sigma_n^2)$, then by the stability property the sum has still a normal distribution $N(m_1 + ... + m_n, \sigma_1^2 + ...\sigma_n^2)$.

It is also the case for the Poisson and the compound Poisson distributions defined below.

PROPERTY 6.1.– [APP 04] If $L = (L(t), t \geq 0)$ is a Lévy process, then for all t, L(t) is infinitely divisible.

PROOF.– Let us define for all n the following random variables:

$$Y_k^{(n)} = L(\frac{kt}{n}) - L(\frac{(k-1)t}{n}), k = 1,...,n \qquad [6.6]$$

so that:

$$L(t) = Y_1^{(n)}(t) + ... + Y_n^{(n)}(t).$$

From relation [6.3] of definition 6.1, the r.v. $Y_1^{(n)},...,Y_n^{(n)}$ are i.i.d., $Y_1^{(n)},...,Y_n^{(n)}$, *X(t)* is infinitely divisible. □

From this property, it follows that the c.f. ϕ the increments of the process L satisfies the following property:

$$\forall s,t > 0: \ \phi_{L_{t+s}-L_s}(u) = [\phi(u)]^t. \qquad [6.7]$$

In particular, all the increments of type $L_{t+1} - L_t$, and, in particular, L_1, have *ϕ(u)* as c.f. and thus the same d.f. F corresponding to this c.f. With $s = 0$ in this last relation, we see that

$$\forall t > 0: \ \phi_{L_t}(u) = [\phi(u)]^t. \qquad [6.8]$$

If we define a Lévy process by the d.f. of L_t, it suffices to choose the d.f. of one of these L_t variables, for example L_1.

PROPOSITION 6.1 [SAT 99].– Let F be an infinitely divisible distribution. Then, there exists a Lévy process $L = (L_t)$ such that the distribution of L_1 is given by F.

PROPERTY 6.2 [MIK 09].– If L is a Lévy process, then

i) If there exists some s>0 such that $\text{var}[L(s)]$ exists, then $\text{var}[L(1)]$ exits and for all positive t $\text{var}[L(t)] = t\,\text{var}[L(1)]$.

ii) If there exists some s>0 such that $E[L(s)]$ exists, then $E[L(1)]$ exits and for all positive t $E[L(t)] = tE[L(1)]$.

PROPERTY 6.3.– Let L_1, L_2 be two independent Lévy processes; then any linear combination $aL_1 + bL_2, a, b \in \mathbb{R}$ is a Lévy process.

The proof of this property is simple as the five conditions of definition 6.1 remain still valid because both L_1, L_2 remain true for the linear combination.

PROPERTY 6.4 [PRO 90].– Every Lévy process has a unique càdlàg modification.

Therefore, we will henceforth always assume that the considered Lévy process is càdlàg that is almost surely continuous from the right and having a left limit for every *t*.

6.3. Lévy–Khintchine formula

For any Lévy process *L*, let us now introduce the generating function of cumulants $\psi(u)$, which is a continuous function, defined as follows:

$$\psi(u) = \ln \phi(u) \qquad [6.9]$$

where we know that

$$\phi(u) = E\left[e^{iuL_1}\right] \qquad [6.10]$$

Therefore:

$$\phi(u) = e^{\psi(u)}. \tag{6.11}$$

With this notation, the complex function $\psi(u)$ is often called the *characteristic exponent* or *Lévy exponent*.

The *Lévy–Khintchine theorem* [BRE 68] asserts that the Lévy exponent $\psi(u)$ satisfies the following relation:

$$\psi(u) = i\gamma u - \frac{1}{2}\sigma^2 u^2 + \int_{-\infty}^{+\infty} (e^{iux} - 1 - iux1_{\{|x|<1\}})\nu(dx), \tag{6.12}$$

where γ and σ are two real constants, and ν is called a *Lévy measure*, that is a measure on $\mathbb{R}_0$ such that:

$$\int_{\mathbb{R}_0} \inf\{1, x^2\}\nu(dx) < \infty. \tag{6.13}$$

REMARK 6.1.– Let us note that this last condition implies that for all θ > 0, we have

$\nu(\{x : |x| > \theta\} < \infty$, but there may exist a θ such that $\nu([-\theta, +\theta]) = \infty$.

Indeed, as for $0 < \theta \le 1$, we have $x^2 \wedge \theta \le x^2 \wedge 1$, we have:

$$\nu(\{x : |x| > \theta\} \le \int_{\{x:|x|>\theta\}} x^2 \wedge \theta\nu(dx) < \infty,$$

the last inequality coming from relation [6.13].

To understand the Lévy–Khintchine formula [6.10], we know from definition 6.1 that:

$$E\left[e^{iuL_t}\right] = [\phi(u)]^t \tag{6.14}$$

Therefore, by relation [6.11] giving ϕ as a function of ψ, we get:

$$E\left[e^{iuL_t}\right] = e^{\psi(u)t},$$

and consequently from relation [6.12]:

$$\phi_{L_t}(u) = e^{i\gamma ut - \frac{1}{2}\sigma^2 u^2 t + t\int_{-\infty}^{+\infty}(e^{iux} - 1 - iux 1_{\{|x|<1\}})\nu(dx)}.$$

It follows that ϕ_{L_t} is the product of three characteristic functions:

with:

$$\phi_{L_t}(u) = e^{i\gamma ut} e^{-\frac{1}{2}\sigma^2 u^2 t} e^{it\int_{-\infty}^{+\infty}(e^{iux} - 1 - iux 1_{\{|x|<1\}})\nu(dx)},$$

or

$$\phi_{L_t}(u) = \phi_{1,t}(u)\phi_{2,t}(u)\phi_{3,t}(u) \qquad [6.15]$$

with

$$\phi_{1,t}(u) = e^{i\gamma ut}, \phi_{2,t}(u) = e^{-\frac{1}{2}\sigma^2 u^2 t}, \phi_{3,t}(u) = e^{it\int_{-\infty}^{+\infty}(e^{iux} - 1 - iux 1_{\{|x|<1\}})\nu(dx)}$$

Consequently, the process L is the sum of three processes:

$$L_t = \gamma t + \sigma W_t + J_t. \qquad [6.16]$$

The first process gives the *drift*, the second process gives the standard *Brownian component* independent of the third process, the last one being a *jump process*. This decomposition is called the *Lévy–Itô decomposition* of the Lévy process L.

It follows that a way to define a Lévy process is to give the following triplet: $(\gamma, \sigma, \upsilon)$.

REMARK 6.2.–

1) Of course, if the Lévy measure ν is identically null, this particular case of a Lévy process gives the lognormality model of assets or the Black, Scholes and Samuelson model. Therefore, it is essentially with a non-identically null Lévy measure that we obtain new models for the evolution of a financial asset.

2) Let us mention that the first model with jumps was introduced in [MER 73] in stochastic finance and that Janssen and Manca [JAN 99b] extended this model with the introduction of semi-Markov modelization.

3) If $\sigma = 0$ and $\int_{-1}^{+1} |x| \nu(dx) < \infty$, it can be shown that the considered Lévy process is of bounded variation and inversely if $\sigma = 0$ and $\int_{-1}^{+1} |x| \nu(dx) = \infty$, it is of infinite variation.

The case of bounded variation is not interesting in finance because for the Black and Scholes model, the process of the logarithms of the asset values is already of infinite variation.

1) If $\sigma = 0$ and $\int_{-1}^{+1} \nu(dx) < \infty$, it can be shown that the Lévy process has a finite number of jumps on every finite time interval.

2) From relation [6.12], it is clear that it is the third process jump component $J = (J_t, t \geq 0)$ characterized by

$$\phi_{3,t}(u) = e^{it\int_{-\infty}^{+\infty}(e^{iux}-1-iux1_{\{|x|<1\}})\nu(dx)} \qquad [6.17]$$

that brings a generalization of the Brownian motion.

Now, we can come back to the moments of a Lévy process as a complement to property 6.2.

PROPERTY 6.5.– Let L be a Lévy process of triplet $(\gamma, \sigma, \upsilon)$; then the moment $E\left[\left|L_t^n\right|\right]$ is finite for every t iff $\int_{|x|\geq 1} \left|x^n\right| \nu(dx) < \infty$ and

$$E[L_t] = t(\gamma + \int_{|x|\geq 1} x\nu(dx), \text{var}(L_t) = t(\sigma^2 + \int_{-\infty}^{+\infty} x^2 \nu(dx). \qquad [6.18]$$

Moreover, $E\left[e^{uL_t}\right] < \infty$ for every t iff $\int_{|x|\geq 1} e^{ux} \nu(dx) < \infty$ and

$$E\left[e^{uL_t}\right] = e^{t\psi(-iu)}. \qquad [6.19]$$

6.4. Subordinators

DEFINITION 6.2.– *The process X is called a subordinator if it is, a.s., a non-negative and non-decreasing Lévy process.*

This means that we have a.s.:

$$\begin{aligned}&\forall t, X(t) \geq 0,\\ &\forall s,t \text{ with } s \leq t \Rightarrow X(s) \leq X(t).\end{aligned} \qquad [6.20]$$

Bertoin [BER 99] has proved the following theorem.

PROPERTY 6.6.– The process X is a subordinator iff

$$\psi(u) = ibu + \int_0^{\infty} (e^{iuy} - 1)\lambda(dy)$$

with $b \geq 0, \lambda((-\infty,0))$ and $\int_0^{\infty} \min(y,1)\lambda(dy) < \infty.$ [6.21]

6.5. Poisson measure for jumps

6.5.1. *The Poisson random measure*

Let $N = (N_t, t \geq 0)$ be a Poisson process of parameter λ defined in section 2.7.

We know that the r.v. N_t counts the total number of jumps in $(0,t]$. More generally, let us consider a Borel set U of $\mathbb{R}^+$, and for any sample path ω let us now count the total numbers of jumps in U denoted by:

$$M(\omega, A) = card\{i \geq 1 : T_i \in A\} \qquad [6.22]$$

where from relation [2.8]: $T_0 = 0, T_n = X_1 + \cdots + X_n, n \geq 1$.

Clearly, for each ω, relation [6.21] defines a measure on $[0,\infty)$ such that *M(U)* is a.s. finite for every bounded Borel set A. As this measure depends on ω, it is called a *random measure* or more precisely a *Poisson random measure.*

Here, the mean value of the measure U is given by

$$E[M(U)] = \lambda\mu_L(U) \qquad [6.23]$$

where μ_L is the Lebesgue measure on $(\mathbb{R}, \beta)$.

We also have:

$$N_t(\omega) = M(\omega,[0,t]) = \int_{[0,t]} M(\omega, ds). \qquad [6.24]$$

In summary, the Poisson measure M of parameter λ satisfies the following properties in an application from $\Omega \times \beta \mapsto \mathbb{N}$ such that:

1) for any $A \in \beta$, *M(A)* counts the number of jumps of the Poisson process in the set *A*;

2) if *A* and *B* are two disjoint Borel sets, then the r.v. *M(A)* and *M(B)* are independent;

3) for any Borel set *A*, the r.v. *M(A)* has a Poisson distribution of parameter $\lambda\mu_L(A)$.

6.5.2. *The compensated Poisson process*

The compensated Poisson process is defined as follows:

$$N_t^P = N_t - \lambda t, t \geq 0. \qquad [6.25]$$

The subtraction of the mean of N_t implies that the new process has no trend or drift as $E\left[N_t^P\right] = 0, t \geq 0$.

The Poisson random measure $\tilde{M}$ is defined as:

$$\tilde{M}(\omega, A) = M(\omega, A) - \lambda\mu_L(A) \qquad [6.26]$$

so that now:

$$E\left[\tilde{M}(A)\right] = 0, \text{var}\left[\tilde{M}(A)\right] = \lambda\mu_L(A).$$

Let us remark that this measure is not a positive measure and with values in $\mathbb{R}$ instead with values in $\mathbb{N}$ as for the measure *M*.

6.5.3. *Jump measure of a Lévy process*

Let us measure the *jump at time t* of a Lévy process by:

$$\Delta L_t = L_t - L_{t-},$$

where [6.27]

$$L_{t-} = \lim_{s \nearrow t} L_s.$$

Now let us consider the time interval $[0,t]$ and U a Borel set of $\mathbb{R}$.

The r.v. given the number of jumps on $[0,t]$ with values in U, denoted $N_L(t,U)$, is given by:

$$N_L(t,U) = \sum_{0 \le s \le t} I_U(\Delta L_s)$$

where $\mathrm{I}_U(x) = \begin{cases} 1, x \in U, \\ 0, x \notin U, \end{cases}$ (indicator of the set U). [6.28]

More generally, for any set B of $\mathbb{R} \times [0,\infty)$, we define the counting measure on $\mathbb{R} \times [0,\infty)$ by:

$$J_L(t,B) = card\{(t,(L_t - L_{t-}) \in B\} \quad [6.29]$$

and so:

$$N_L(t,U) = J_L([0,t] \times U). \quad [6.30]$$

PROPERTY 6.7.– The *Lévy measure* ν of the Lévy process L is given by:

$$\forall A \in \beta : \nu(A) = \mu_L \{card\{t \in [0,1] : \Delta L_t \neq 0, \Delta L_t \in A\}\}. \quad [6.31]$$

It represents the expected number of jumps belonging to A per unit of time.

6.5.4. *The Itô–Lévy decomposition*

The Itô–Lévy decomposition [6.16] has another interpretation: any Lévy process can be decomposed as follows [APP 04]:

$$L_t = \gamma t + \sigma B(t) + \int_{|x|<1} x\tilde{N}(t,dx) + \int_{|x|\geq 1} xN(t,dx).$$

where

- γ, σ are constants;
- B is an SBM;
- N is a Poisson random measure on $\mathbb{R}^+ \times \mathbb{R}_0$;
- $\tilde{N}$ is the compensated Poisson random measure of N given by

$\tilde{N}(t,A) = N(t,A) - t\mu(A)$.

Moreover, the process $\left(\int_{|x|<1} x\tilde{N}(t,dx), t \geq 0 \right)$ is the compensated process of small jumps and an L^2-martingale and the process $\left(\int_{|x|\geq 1} xN(t,dx), t \geq 0 \right)$ describing the large jumps is a compound Poisson process.

6.6. Markov and martingale properties of Lévy processes

6.6.1. *Markov property*

Let L be a Lévy process on $(\Omega, \Im, (\Im_t), P)$, where $(\Im_t)$ is the natural filtration defined by $\Im_t = \sigma(L_s, s \leq t)$, that is $\Im_t$ is generated by all the past of the process up to time t.

From relation [6.3], we have:

$$\begin{aligned} &\forall s,t \ (0 \leq s < t), \forall \mathrm{A} \in \Im_t : \\ &P\left[L_t \in A \middle| \Im_s\right] = \left[L_t - L_s \in A - L_s \middle| \Im_S\right] = P\left[L_{t-s} \in A - L_s\right] \end{aligned} \quad [6.32]$$

or

$$P\left[L_t \in A \middle| L_s = x\right] = P\left[L_{t-s} \in A - x\right].$$

Now if we introduce the *transition kernel* of L defined by:

$$P_{s,t}(x,A)=P\left[L_t\in A\middle|L_s\right] \qquad [6.33]$$

the last relation of [6.32] shows that:

$$P_{s,t}(x,A)=P_{0,t-s}(0,A-x)\,. \qquad [6.34]$$

This means that the Lévy process L is a continuous-time Markov process homogeneous in time and space.

In particular, we have:

$$P_{s,t}(x,y)=P(L_t\le y\,|\,L_s=x)=P(L_{t-s}\le y-x). \qquad [6.35]$$

Moreover, if t-s is an integer, say n, we can write:

$$P_{s,t}(x,y)=P(L_n\le y-x). \qquad [6.36]$$

As from relation [6.13], we have:

$$E\left[e^{iuL_n}\right]=[\phi(u)]^n, \qquad [6.37]$$

the distribution of L_n can be seen as the one of a sum of n independent r.v. having

F_{L_1} as distribution function so that:

$$P_{s,s+n}(x,y)=F_{L_1}^{(n)}(y-x), \qquad [6.38]$$

where for any r.v. X, $F_X^{(n)}$ represents the nth product of convolution of F_X so that:

$$F_X^{(2)}(x)=\int_{\mathbb{R}}F(x-s)dF(s),$$

and for n>2: [6.39]

$$F_X^{(n)}(x)=\int_{\mathbb{R}}F^{(n-1)}(x-s)dF(s).$$

Result [6.37] is important if we want to discretize in time a Lévy process.

6.6.2. *Martingale properties*

Let us remark that any Lévy process being a process with independent increments, it has the martingale properties of these processes. This gives the following property.

PROPERTY 6.8 [CON 04].–

Let L be a Lévy process, then:

i) for every real u, the process $\left(\dfrac{e^{iuL_t}}{\phi_{L_t}(u)}, t \geq 0\right)$ is a martingale;

ii) if $E\left[\left|L_1\right|\right] < \infty$, then the process $(M_t, t \geq 0)$ where $M_t = L_t - E\left[L_t\right]$ is a martingale and with independent increments;

iii) if $\operatorname{var}\left[L_1\right] < \infty$, then the process $\left((\mathrm{M}_t^2 - (E\left[M_t\right])^2, \mathrm{t} \geq 0)\right)$ is a martingale;

iv) in the special case of a Brownian motion, we obtain properties 5.2 and 5.3.

REMARK 6.3.–

As from relation [6.11], we have $\phi(u) = e^{\psi(u)}$, the first part of property [6.8] affirms that the process $\left(\mathrm{e}^{iuL_t - t\psi(u)}, t \geq 0\right)$ is a martingale.

The next property gives necessary and sufficient conditions.

PROPERTY 6.9 [CON 04].–

i) The Lévy process is a martingale iff:

$$\gamma + \int_{|x| \geq 1} x\nu(dx) = 0, \qquad [6.40]$$

ii) The process $\left(e^{L_t}, t \geq 0\right)$ is a martingale iff:

$$\begin{aligned} &\int_{|x| \geq 1} e^x \nu(dx) < \infty, \\ &\frac{\sigma^2}{2} + \int_{-\infty}^{+\infty} (e^x - 1 - x 1_{\{|x \geq 1|\}}) \nu(dx) = 0. \end{aligned} \qquad [6.41]$$

The next property is very important for the applications as it will give the right to use any levy process as integrator process in the sense of Chapter 5 where we said that we can define the stochastic integral on $[0,t]$ *for every adapted càglàd process as integrand process and for every semi-martingale integrator process.*

PROPERTY 6.10 [APP 04].– Every Lévy process is a semi-martingale.

It follows that we can use the Itô integral with Lévy processes as integrators and also the Itô formula.

6.6.3. *Itô formula*

In section 5.5.2, we gave the general form of Itô formula, and here we can use it for Lévy processes.

The Lévy process L will be considered as integrator for an integrand process X to give the following stochastic integral:

$$X_t(\omega) = \int_0^t Y_s(\omega)dL_s(\omega)(X(t) = \int_0^t Y_s dL_s \text{ in short}). \qquad [6.42]$$

Itô formula: If X is given by relation [6.42] and f is a function of class $C^2_{\mathbb{R}}$, then the composed process $f(X) = (f(X_t), t \geq 0)$ is given as:

$$\begin{aligned} f(X_t) - f(X_0) = &\int_{0+}^t f'(X_{s-})dX_s + \frac{1}{2}\int_{0+}^t f'(X_{s-})d[X,X]_s^c \\ &+ \sum_{0<s\leq t} \{f(X_s) - f(X_{s-}) - f'(X_s)\Delta X_s\}. \end{aligned} \qquad [6.43]$$

6.7. Examples of Lévy processes

6.7.1. *The lognormal process: Black and Scholes process*

From the results of Chapter 5 and remark 6.2, we see that this model corresponds to the particular Lévy process of triplet $(\mu - \frac{\sigma^2}{2}, \sigma, 0)$ as indeed in this case we have that the r.v. L_1 has a normal distribution of parameters $(\mu - \frac{\sigma^2}{2}, \sigma, 0)$.

6.7.2. *The Poisson process*

For a Poisson law of parameter λ, we know that (see section 6.1):

$$\phi_\lambda(u) = e^{\lambda(e^{iu}-1)}.$$

From the stability property of the Poisson law, if X and Y are two independent r.v. of parameters λ_1, λ_2, the sum also follows a Poisson law of parameter $\lambda_1 + \lambda_2$ and has the c.f.

$$\phi_{\lambda_1+\lambda_2}(u)(= e^{\lambda_1+\lambda_2(e^{iu}-1)}).$$

Consequently, the Poisson law is infinitely divisible and:

$$\psi_\lambda(u) = \ln \phi_\lambda(u) = \lambda(e^{iu}-1). \qquad [6.44]$$

Now, the *Poisson process of parameter* $\lambda, (N_t, t \geq 0)$, is defined as a pure jump process with all jumps of amplitude 1, jumps occurring at times $X_1, X_1 + X_2, ..., X_1 + X_2 + ... + X_n, ...$, where the r.v. $(X_n, n \geq 1)$ are i.i.d. having an exponential distribution of parameter λ.

It follows that the r.v., given the total number of jumps in (0.*t*), has a Poisson distribution of parameter λt:

$$P(N_t = k) = e^{-\lambda t}\frac{(\lambda t)^k}{k!}, k = 0,1,... \qquad [6.45]$$

A possible sample path of this process is given in Figure 6.1.

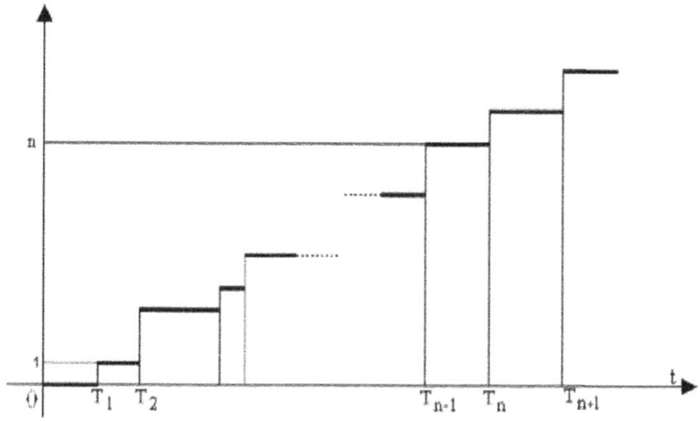

Figure 6.1. *A possible sample path of the process N*

Here, we have:

$$\psi_\lambda(u) = \ln \phi_\lambda(u) = \lambda(e^{iu} - 1) . \qquad [6.46]$$

So, this process is a particular Lévy process of triplet:

$$\{0, 0, \lambda\delta(1)\},$$

(1) being the Dirac measure with a unit mass at $x = 1$.

Clearly, the Poisson process is a subordinator.

6.7.3. *Compensated Poisson process*

Let us consider the compensated Poisson process given in section 6.5.2 defined by:

$$M_t^P = N_t - \lambda t, t \geq 0.$$

We know that the subtraction of the mean of N_t implies that the new process has no trend or drift as $E\left[M_t^P\right] = 0, t \geq 0.$

From Table 6.1, we have:

$$\phi_N(t) = \exp(\lambda t(e^{iu} - 1))$$

and so:

$$\begin{aligned}\phi_M(t) &= \exp(\lambda t(e^{iu} - 1)) * \exp(-i\lambda tu), \\ &= \exp(\lambda t(e^{iu} - 1 - iu).\end{aligned}$$

6.7.4. *The compound Poisson process*

To define this process, let us consider a Poisson process N of parameter and a sequence $(Y_n, n \geq 1)$ of non-negative i.i.d. with B as the d. f. and mean β.

The compound Poisson process $U = (U_t, t \geq 0)$ is defined as follows:

$$U_t = \begin{cases} 0, & N_t = 0, \\ \sum_{n=1}^{N(t)} Y_n, N_t > 0. \end{cases} \qquad [6.47]$$

Its meaning in the actuarial theory of risk is as follows: let us consider an insurance company for which the process of claim arrivals is a Poisson process of parameter λ and the successive claim amounts are given by the r.v. $Y_1,...,Y_n,...$. If u represents the equities amount at t = 0 and if the premium rate p per time unit is given by the constant c, then the net asset of the insurer at time t is given by:

$$\alpha_t = u + ct - \sum_{n=1}^{N_t} Y_n, \qquad [6.48]$$

provided that the ruin event, that is a first passage toward a negative net asset before t, has not occurred. A sample path of this process with the occurrence of the ruin event is shown in Figure 6.1, where:

$$T_1 = X_1, T_2 = X_1 + X_2, ..., T_n = X_1 + ... + X_n, ... \qquad [6.49]$$

and r.v. X_n represents the interarrival times between two consecutive claims exponentially distributed with parameter λ.

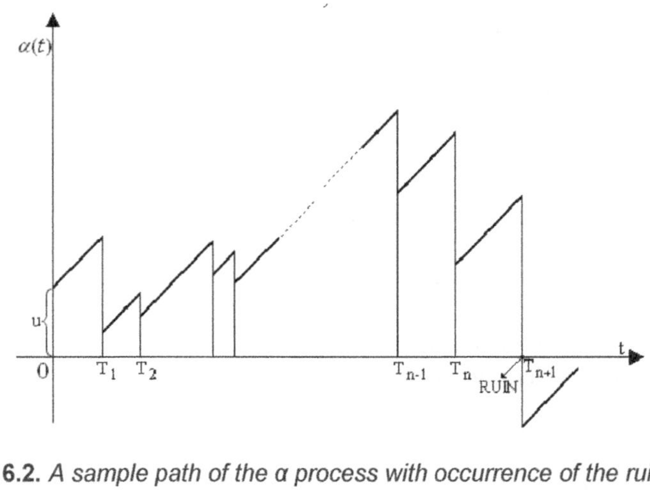

Figure 6.2. *A sample path of the α process with occurrence of the ruin event*

It is possible (see, e.g. [JAN 07]) to prove that the c.f. of U_t is given by:

$$\phi_{U_t}(u) = e^{t\lambda(\phi_Y(u)-1)}. \qquad [6.50]$$

It follows:

$$\psi_{U_t}(u) = t\lambda(\phi_Y(u) - 1). \qquad [6.51]$$

Now, let us introduce the measure defined as follows:

$$\nu_P(A) = \begin{cases} 0, & A = \{0\}, \\ \lambda P(Y \in A), & A \in \beta - \{0\}. \end{cases} \tag{6.52}$$

Then, it is possible to show that:

$$\phi_{U_t}(u) = e^{t\int_{-\infty}^{+\infty}(e^{iux}-1)\nu_P(dx)}, \tag{6.53}$$

and therefore the compound Poisson process is a Lévy process of triplet $\left\{\int_{-1}^{+1} x\nu_P(dx), 0, \nu_P\right\}$.

Clearly, the compound Poisson process is a subordinator.

It can be proved (see [CON 04]) that U is a compound process iff it is a Lévy process with piecewise constant sample paths.

Let us note that this process will be used in Volume 2 of the series devoted to non-life insurance.

6.8. Variance gamma (VG) process

6.8.1. *The gamma distribution*

Let us recall that for a *gamma distribution of parameters a and b*, we know (see section 1.3.6) that its density function is given by:

$$f(x) = \begin{cases} 0, x < 0, \\ \dfrac{b^a}{\Gamma(a)} x^{a-1} e^{-xb}, x \geq 0 \end{cases}$$

with, from section 6.1:

$$\phi_X(t) = \frac{1}{(1 - \frac{it}{b})^a} \tag{6.54}$$

So the gamma distribution is infinitely divisible.

6.8.1.1. *The gamma process*

The process $X^G = \left(X_t^G, t \geq 0\right)$ is a *gamma process* if it starts from 0 and has stationary and independent increments such that the distribution of X_1^G is a gamma distribution of parameters a and b.

It follows that the distribution of X_t^G is a gamma distribution of parameters at, b.

The process $X^G = \left(X_t^G, t \geq 0\right)$ is a Lévy process as it is an infinitely divisible process. Then it can be shown computing relations [6.11] and [6.12] that the triplet of $X^G = \left(X_t^G, t \geq 0\right)$ is:

$$\left(\frac{a}{b}(1-e^{-b}), 0, \frac{a}{x}e^{-bx}1_{\{x>0\}}\,dx\right). \qquad [6.55]$$

6.8.2. *The VG distribution*

An r.v. Z has a *VG distribution of parameters C, G and M* (see [MAD 90]) if it is the difference of two independent r.v. X and Y, each of them having a gamma distribution, respectively, with parameters (C, M) and (C, G):

$$Z = X - Y \prec VG(C, G, M).$$

It can be shown that [MAD 98]:

$$\begin{aligned} \phi_Z(t) &= \left(\frac{GM}{GM+(M-G)it+t^2}\right)^C \\ f_Z(x) &= \frac{(GM)^C}{\sqrt{\pi}\Gamma(C)}e^{\frac{(G-M)x}{2}}\left(\frac{|x|}{G+M}\right)^{C-1/2} K_{C-1/2}\left(\frac{(G+M)|x|}{2}\right), \end{aligned} \qquad [6.56]$$

with the function K_ν representing the Bessel function of order 3 with the index ν.

A computation [MAD 98] shows that:

$$\begin{aligned}
E[Z] &= \frac{C(G-M)}{MG},\\
\operatorname{var} Z &= \frac{C(G^2+M^2)}{(MG)^2},\\
\text{skewness} &= \frac{2(G^3-M^3)}{\sqrt{C}(G^2+M^2)^{3/2}},\\
\text{kurtosis} &= \frac{3[1+2C^{-1}(G^4+M^4)]}{(G^2+M^2)^2}.
\end{aligned} \qquad [6.57]$$

In particular, the VG distribution is symmetrical for $G = M$. It has a negative kurtosis coefficient if $G < M$ and a positive coefficient if $G > M$.

6.8.3. *The VG process*

By definition, the stochastic process $X = (X_t, t \geq 0)$ is called a *VG process* of parameters C, G and M, all strictly positive, denoted by VG (C, G, M) if:

i) X has independent increments;

ii) X has stationary increments;

iii) $\forall\, s,t > 0 : X_{t+s} - X_t \prec VG(Cs, G, M)$.

It can be shown that this process has no Brownian component, i.e. $\sigma = 0$, and that:

$$\begin{aligned}
\gamma &= \frac{-C(Ge^{-M}-1) - M(e^{-G}-1)}{MG},\\
\nu_{VG}(dx) &= \begin{cases} Ce^{Gx}|x|^{-1}\,dx, x<0,\\ Ce^{-Mx}|x|^{-1}\,dx, x>0. \end{cases}
\end{aligned} \qquad [6.58]$$

This Lévy measure has an infinite mass and so has infinitely many jumps in any finite interval, of course with infinitesimal amplitudes in every finite time interval, and moreover all the sample paths are of finite variation.

From relations [6.55] and [6.58], we have the following decomposition:

$$X_t^{VG} = \left(X_t^{G_1}\right) - X_t^{G_2}$$

where:

$$\left(X_t^{G_1}\right) \text{is } G(C,M) \text{ and } \left(X_t^{G_2}\right) \text{is } G(C,G).$$

We also have:

$$E\left[X_1^{VG}\right]=\phi_{VG}(u,C,G,M)=\left(\frac{GM}{GM+(M-G)iu+u^2}\right)^C \qquad [6.59]$$

6.8.4. *The Esscher transformation*

On our basic probability space $(\Omega,\Im,P)$, let *f(t,x)* be the density function of X_t. If θ represents a real number such that $E\left[e^{\theta X_t}\right]<\infty$, we can introduce a new density $f^{(\theta)}$ defined by:

$$f^{(\theta)}(x,t)=\frac{e^{\theta x}f(x,t)}{\int_{-\infty}^{+\infty}e^{\theta y}f(y,t)dy}. \qquad [6.60]$$

In non-Gaussian stochastic finance, the VG process is used to model an asset value as follows:

$$S_t=S_0e^{X_t} \qquad [6.61]$$

where the process of logarithms of the ratio of two consecutive asset values

$$\ln\frac{S_{t+1}}{S_t}=X_{t+1}-X_t \qquad [6.62]$$

is supposed to be a VG (*C, G, M*) replacing the lognormal distribution in the Black and Scholes model.

As for the Black and Scholes model, we should like to pass from the historical measure *P* to the so-called risk neutral measure *Q* for which the process of present values of the asset values $(\hat{S}_t,t\in[0,T])=S({}_0e^{(X(t)-rt)},t\in[0,T])$ is a martingale, *r* being as usual the no-risky instantaneous interest rate.

Therefore, if $P^{(\theta)}$ represents the corresponding measure to the density defined by relation [6.60], we must find the θ value(s) such that we have:

$$S_0=e^{-rt}E_{P^{(\theta)}}[S_t],t\in[0,T]. \qquad [6.63]$$

It can be shown that this equality leads to the following algebraic equation:

$$e^r = \frac{\phi(-i(\theta+1)}{\phi(-i\theta)}. \qquad [6.64]$$

If θ^* is a solution, we obtain the risk-neutral measure $Q = Q^{(\theta^*)}$ that we can utilize to compute the value of a European call with K as exercise price and of maturity T with the following formula:

$$C(0,T) = \int_{\ln\frac{K}{S_0}}^{\infty} f^{(\theta^*+1)}(x,T)dx - e^{-rT}\int_{\ln\frac{K}{S_0}}^{\infty} f^{(\theta^*)}(x,T)dx. \qquad [6.65]$$

EXAMPLE 6.1.–

Let us consider the Black and Scholes model presented in section 6.7 for which we have seen that the corresponding Lévy process has triplet $(\mu-\frac{\sigma^2}{2},\sigma,0)$ and so:

$$\phi(u) = e^{\left[iu(\mu-\frac{\sigma^2}{2})-\frac{\sigma^2 u^2}{2})\right]}.$$

Equation [6.64] becomes:

$$r = (\theta+1)(\mu-\frac{\sigma^2}{2}) - \frac{\sigma^2(\theta+1)^2}{2} - \theta(\mu-\frac{\sigma^2}{2}) + \frac{\sigma^2\theta^2}{2}$$

and after simplification, we obtain:

$$r = \mu - \frac{\sigma^2}{2} + \frac{\sigma^2(2\theta+1)}{2}$$

and so:

$$\theta^* = \frac{r-\mu}{\sigma^2}.$$

Now, using result [6.60] with the value found for θ^*, we obtain:

$$f_1^{\theta^*}(x) = \frac{\exp(\theta^* x - \frac{(x-\mu+\frac{1}{2}\sigma^2)^2}{2\sigma^2})}{\int_{-\infty}^{+\infty}\exp(\theta^* y - \frac{(y-\mu+\frac{1}{2}\sigma^2)^2}{2\sigma^2})dy}.$$

Computing this expression, we obtain well the result that $f_1^{\theta*}(x)$ is the density of a normal distribution of parameters $(r-\frac{\sigma^2}{2},\sigma^2)$.

REMARK 6.4.–

The Esscher transformation is not the only possibility to get a risk-neutral measure. For example, there also exists the *method of correcting the exponential* of a Lévy process, i.e. by introducing a new Lévy process $\tilde{X}$ defined by:

$$\tilde{X}_t = X_t + mt \tag{6.66}$$

It is possible to compute m so that the process of present values of the asset values

$$(\hat{S}_t, t \in [0,T]) = S(_0 e^{(X(t)-rt)}, t \in [0,T])$$

is a martingale (see [SCH 03]).

It follows that for incomplete markets we have many ways to obtain the martingale measure and the only way to see whether we have the good one for the considered market is to calibrate on this market.

From a theoretical point of view, some authors give arguments for the Esscher transformation (see [GER 96]).

6.8.5. *The Carr–Madan formula for the European call*

Using the method of correcting the exponential explained in remark 6.4, the process of present values defined by

$$\tilde{S}_t = S_0 e^{(r-\beta^*)t+X_t}$$

of the asset values becomes a martingale under the risk-neutral measure $Q^{(\beta^*)}$ with:

$$\beta^* = C\ln\left[\frac{(M-1)(G+1)}{MG}\right]. \tag{6.67}$$

So, we have:

$$C(K,T) = e^{-rT} E_{Q^{(\beta^*)}} \left[\max\{S_T - K, 0\} \right].$$

As the distribution of S_T is a known VG, it is theoretically possible to compute this value. However, the result is complicated and difficult to be applied.

Nevertheless, Carr and Madan [CAR 99] obtained such a result using the Fourier transform introducing a parameter α such that the function $e^{\alpha k}C(k,T)$ is a square integrable:

$$C(K,T) = \frac{e^{-\alpha \ln K}}{\pi} \int_0^{\infty} e^{-is\ln K} \frac{e^{-rT}\phi(s-(\alpha+1)i)\ln S_T}{\alpha^2+\alpha-s^2+i(2\alpha+1)s} ds.$$

Several authors [CAR 99, SCH 03] computed the call value using the fast Fourier transform (FFT) with a discretization of this last result.

It is clear that such results are not operational particularly in market rooms.

6.9. Hyperbolic Lévy processes

In physics, Barndorff-Nielsen [BAR 77a] introduced a distribution with four real parameters $\alpha, \beta(\alpha > \beta), \delta, \mu$ defined by the following probability density function:

$$h_{\delta,\mu}^{\alpha,\beta}(x) = \frac{\sqrt{\alpha^2-\beta^2}}{2\alpha\delta K_1\left(\delta\sqrt{\alpha^2-\beta^2}\right)} \exp\left[-\alpha\sqrt{\delta^2+(x-\mu)^2}+\beta(x-\mu)\right], x \in \mathbb{R} \qquad [6.68]$$

where K_1 is the Bessel function of the first kind of order 1. The Bessel functions of the first kind of order v are defined by:

$$K_\nu(x) = \frac{1}{2}\int_0^{\infty} u^{\nu-1} \exp\left[-\frac{1}{2}x(u+\frac{1}{u})\right] du, x \in \mathbb{R}. \qquad [6.69]$$

This distribution is called hyperbolic *distribution* as $\ln h_{\delta,\mu}^{\alpha,\beta}(x)$ is a hyperbolic function.

REMARK 6.5 (Special cases of hyperbolic distribution).–

1) If $\mu = 0$, $\delta = 0$, [6.68] gives the density function of the variance gamma distribution.

2) For $\alpha = \beta = \mu = 0, \lambda < 0$, [6.68] gives the density function of Student's distribution.

3) For $\lambda = -\frac{1}{2}$, [6.68] gives the density function of the normal inverse distribution.

In [BAR 77a, BAR 77b] and [EBE 98], we can find the proofs of the main properties of these distributions given in the following proposition.

PROPOSITION 6.2.–

i) Every hyperbolic distribution is infinitely divisible.

ii) Any hyperbolic distribution has moments of any order.

iii) For $|u+\beta<\alpha,|$ the *generating function* exists and is given by:

$$M_{\delta,\mu}^{\alpha,\beta}(u) = e^{\mu u} \frac{\sqrt{\alpha^2-\beta^2}}{K_1(\delta\sqrt{\alpha^2-\beta^2})} \frac{K_1(\delta\sqrt{\alpha^2-(\beta+u)^2})}{\sqrt{\alpha^2-(\beta+u)^2}}. \qquad [6.70]$$

iv) The mean and variance of an r.v. X with a hyperbolic distribution given by relation [6.57] are given by:

$$E[X] = \mu + \frac{\delta\beta}{\sqrt{\alpha^2-\beta^2}} \frac{K_2(\zeta)}{K_1(\zeta)},$$

$$\mathrm{var}[X] = \delta^2 \left[\frac{K_2(\zeta)}{\zeta K_1(\zeta)} + \frac{\beta^2}{\alpha^2-\beta^2} \left(\frac{K_3(\zeta)}{K_1(\zeta)} - \left(\frac{K_2(\zeta)}{K_1(\zeta)} \right)^2 \right] \qquad [6.71]$$

where $\zeta = \delta\sqrt{\alpha^2-\beta^2}$ and

$$K_\nu(x) = \frac{1}{2}\int_0^\infty y^{\nu-1} e^{-\frac{1}{2}x(y+\frac{1}{y})} dy (\sim \frac{1}{2}\Gamma(\nu)\left(\frac{z}{2}\right)^{-\nu}, \nu > 0.$$

DEFINITION 6.3.– *A Lévy process such that the c.f. ϕ of L_1 is given by property*

$$\phi_{\delta,\mu}^{\alpha,\beta}(u) = e^{i\mu u} \frac{\sqrt{\alpha^2-\beta^2}}{K_1(\delta\sqrt{\alpha^2-\beta^2})} \frac{K_1(\delta\sqrt{\alpha^2-(\beta+iu)^2})}{\sqrt{\alpha^2-(\beta+iu)^2}}. \qquad [6.72]$$

is called a *hyperbolic Lévy process* and is represented by $X_{\delta,\mu}^{\alpha,\beta}$.

REMARK 6.6.– From remark 6.1, the case $\mu = 0,\ \delta = 0$ gives *the variance gamma Lévy process.*

6.10. The Esscher transformation

In this section, we will give another representation of the Esscher transformation already presented in section 6.8.4.

6.10.1. *Definition*

Let us consider a Lévy process L with triple (γ,σ,υ) on $(\Omega,\Im,P)$ such that:

$$\exists\theta\in\mathbb{R}:\int_{|x|\geq 1} e^{\theta x}\nu(dx)<\infty. \qquad [6.73]$$

Let us change the measure ν by replacing it by the measure ν' such that:

$$\frac{d\nu'}{d\nu}=e^{\theta x}\text{ or }\nu'(dx)=e^{\theta x}\nu(dx). \qquad [6.74]$$

This transformation is known as the *Esscher transformation* and it can be shown (see [SAT 99] and [CON 04]) that on the new probability space $(\Omega,\Im,Q)$, the Lévy process L has as new triplet $(\gamma',\sigma',\upsilon')$ with:

$$\sigma'=0,\gamma'=\gamma+\int_{-1}^{+1}x(e^{\theta x}-1)\nu(dx). \qquad [6.75]$$

And the new probability measure is given by its Radon–Nikodym derivative:

$$\frac{dQ}{dP}=\frac{e^{\theta L_t}}{E\left[e^{\theta L_t}\right]} \qquad [6.76]$$

which can be written as:

$$\frac{dQ}{dP}=e^{(\theta L_t+\gamma(\theta)t)}$$

with $\gamma(\theta)=-\ln E\left[e^{\theta t}\right]$ [6.77]

With this result, we can convert each expectation in the measure P into the measure Q and conversely as follows:

$$\begin{aligned} E_P\left[f(X)\right] &= \int f(X(\omega))dP = \int f(X(\omega))e^{-(\theta L_t+\gamma(\theta)t)}dQ, \\ E_Q\left[f(X)\right] &= \int f(X(\omega))dQ = \int f(X(\omega))e^{(\theta L_t+\gamma(\theta)t)}dP. \end{aligned} \qquad [6.78]$$

6.10.2. *Option theory with hyperbolic Lévy processes*

This transformation is very useful in stochastic finance using exponential of a Lévy process on $(\Omega, \Im, P)$ so that the price of the asset at time t is given by $S_t = S_0 e^{L_t}$, where L is a Lévy process.

Let us consider the model used by Eberlein and Keller [EBE 95] for which the asset follows on [0,T] an exponential hyperbolic Lévy process with financial volatility equal to 1 so that:

$$S_t = S_0 e^{X_{\zeta,\delta}(t)} \qquad [6.79]$$

where $X_{\zeta,\delta}$ is a particular $X_{\delta,0}^{\alpha,0}$ Lévy process defined by:

$$X_{\zeta,\delta}(t) = X_{\delta,0}^{\alpha,0}(t) \text{ with } \zeta = \delta\alpha. \qquad [6.80]$$

Using result [6.68], the density function of $X_{\zeta,\delta}(t)$ is given by:

$$f_{\zeta,\delta}(t) = \frac{1}{2\delta K_1(\zeta)} e^{-\zeta\sqrt{1+(\frac{x}{\delta})^2}}. \qquad [6.81]$$

The present value process of S, denoted $\widehat{S}$ with a fixed non-risky instantaneous rate r, is given by:

$$\widehat{S}_t = S_0 e^{(X_{\zeta,\delta}(t)-rt)}, t \in [0,T]. \qquad [6.82]$$

For the pricing of a European option with K as exercise price and maturity T, we know from stochastic finance that we need to find a new measure Q such that the process $\widehat{S}$ is a martingale giving as formula for the call at time 0, $C(S_0,T)$:

$$C(S_0,T) = E_Q\left[\max\{S_T - K, 0\}\right]. \qquad [6.83]$$

It is possible to show that there is a free arbitrage market if and only if such a martingale exists and is unique. In this case, the market is called complete. But here Eberlein and Keller [EBE 95] proved that the market is *incomplete* and that there exist an infinite number of such martingales. Also, they utilized the Esscher transform to find a martingale measure Q_u with Radon–Nikodym measure given by relation [6.77], but here with $\theta = -u$:

$$\begin{aligned} &\frac{dQ_u}{dP} = N_u(t) \\ &\textit{with} \\ &N_u(t) = e^{-uX_{\zeta,\delta}(t) - t\ln \mathrm{M}_{\zeta,\delta}(\mathrm{u})} \\ &\mathrm{M}_{\zeta,\delta}(t) = E_P\left[e^{uX_{\zeta,\delta}(t)}\right]. \end{aligned} \tag{6.84}$$

From relation [6.70], and as from definition 6.69, $\mu = \beta = 0, \delta = \frac{\alpha}{\zeta}$, we know that:

$$\mathrm{M}_{\zeta,\delta}(t)(u) = \frac{\alpha}{K_1(\delta\alpha)} \frac{K_1\left(\frac{\zeta}{\alpha}\sqrt{\alpha^2 - u^2}\right)}{\sqrt{\alpha^2 - u^2}}$$

and as $\alpha = \frac{\zeta}{\delta}$:

$$\begin{aligned} \mathrm{M}_{\zeta,\delta}(t)(u) &= \frac{\zeta}{\delta K_1(\zeta)} \frac{K_1\sqrt{\zeta^2 - \delta^2 u^2}}{\sqrt{(\frac{\zeta}{\delta})^2 - u^2}} \\ &= \frac{\zeta}{K_1(\zeta)} \frac{K_1\sqrt{\zeta^2 - \delta^2 u^2}}{\sqrt{\zeta^2 - \delta^2 u^2}}. \end{aligned} \tag{6.85}$$

As the process $\left(\widehat{S}(t), 0 \le t \le T\right)$, here with $S_0 = 1$, is a Q-martingale iff the process $\left(\widehat{S}(t)N_u(t), 0 \le t \le T\right)$ is a P-martingale where:

$$\widehat{S}(t)N_u(t) = \exp\left[(1-u)X_{\zeta,\delta}(t) - t(\mathrm{r} + \ln \mathrm{M}_{\zeta,\delta}(\mathrm{u})\right]. \tag{6.86}$$

As the process

$$\left(\exp\left[(1-u)X_{\zeta,\delta}(t) - t(\ln\{\mathrm{M}_{\zeta,\delta}(1-\mathrm{u})\})\right], 0 \le t \le T\right) \tag{6.87}$$

is a martingale, we see that the process $\widehat{S}$ is a martingale iff the following condition is satisfied:

$$r = \ln M_{\zeta,\delta}(1-u) - \ln M_{\zeta,\delta}(u). \qquad [6.88]$$

Finally, using the last equality of relation [6.74], we obtain:

$$r = \ln\left(\frac{K_1(\sqrt{\zeta^2 - \delta^2(1-u)^2}}{K_1(\sqrt{\zeta^2 - \delta^2 u^2}}\right) - \frac{1}{2}\ln\left(\frac{\zeta^2 - \delta^2(1-u)^2}{\zeta^2 - \delta^2 u^2}\right). \qquad [6.89]$$

Then, it suffices to use numerical methods, particularly for the computation of Bessel functions to obtain the value(s) u^* of u giving martingale measure(s).

6.10.3. *Value of the European option call*

We know that the value of the European option call with K as exercise price and maturity T is given by:

$$C(0,T) = e^{-rT} E_{Q_{u^*}}\left[(S(T) - \mathrm{K})^+\right] \qquad [6.90]$$

and from relation [6.79]:

$$C(0,T) = e^{-rT} E_{Q_{u^*}}\left[(S_0 e^{X_{\zeta,\delta}(T)} - \mathrm{K})^+\right]. \qquad [6.91]$$

Using the Esscher transform, we see that the density function of $X_{\zeta,\delta}(t)$ under the probability measure Q_{u^*} designed by $f^t_{\zeta,\delta}(x;u)$ is given by:

$$f^t_{\zeta,\delta}(x;u) = f^t_{\zeta,\delta}(x)\exp^{-\left[ux + t\ln M_{\zeta,\delta}(u)\right]}, x \geq 0 \qquad [6.92]$$

where $f^t_{\zeta,\delta}(x)$ is the density function of $X_{\zeta,\delta}(t)$ under the probability measure P given by relation [6.70].

It follows that the final formula to compute the European call is given by:

$$C(0,T) = \int_{\ln\frac{K}{S_0}}^{\infty} f^T_{\zeta,\delta}(x,1-u)dx - Ke^{-rT}\int_{\ln\frac{K}{S_0}}^{\infty} f^T_{\zeta,\delta}(x,u)dx. \qquad [6.93]$$

This formula replaces the Black and Scholes formula only true in the Brownian geometric model.

6.11. The Brownian–Poisson model with jumps

One of the first models with jumps to be used in stochastic finance was proposed by Merton [MER 76] and was extended by El Karoui and Gobet [ELK 01]. We will summarize here the principal results concerning these models.

6.11.1. *Mixed arithmetic Brownian–Poisson and geometric Brownian–Poisson processes*

The mixed *geometric Brownian–Poisson process* model is an extension of the classical Black and Scholes model, where the process S is given by:

$$S(t) = S_0 e^{\left(\mu-\frac{\sigma^2}{2}\right)t+\sigma B(t)+\sum_{n=1}^{N(t)}\ln(1+Y_n)}$$

or [6.94]

$$S(t) = S_0 e^{\left(\mu-\frac{\sigma^2}{2}\right)t+\sigma B(t)}\prod_{n=1}^{N(t)}(1+Y_n), 0 \le t \le T$$

where the process B is an SBM and the process

$$\left(\sum_{n=1}^{N(t)}\ln(1+Y_n), t \ge 0\right) \qquad [6.95]$$

is a compound Poisson process with parameter λ; the r.v. Y_n has F as the common d.f. with $Y_n > -1$ for all n. Moreover, the two processes are independent. Then, the process S is called a mixed *geometric Brownian–Poisson process.*

We can also write:

$$\frac{S(t)}{S_0} = e^{X(t)} \text{ or } \ln\frac{S(t)}{S_0} = X(t) \qquad [6.96]$$

$$X(t) = \left(\mu-\frac{\sigma^2}{2}\right)t+\sigma B(t)+\sum_{n=1}^{N(t)}\ln(1+Y_n).$$

This process is well a Lévy process.

More generally, if $(Z_n, n \geq 1)$ is a sequence of i.i.d. random variables independent of the process B, $x_0, \mu, \sigma(>0)$ constants , the process defined by

$$X(t) = x_0 + \mu t + \sigma B(t) + \sum_{n=1}^{N(t)} Y_n \qquad [6.97]$$

is called a mixed arithmetic Brownian–Poisson process.

PROPOSITION 6.3.– If X is a mixed arithmetic Brownian–Poisson process represented by relation [6.97], then:

$$\begin{aligned} E\big(X(t)\big) &= x_0 + \mu t + \lambda t E[Y], \\ \operatorname{var}\big(X(t)\big) &= \sigma^2 t + \lambda t E\left[Y^2\right]. \end{aligned} \qquad [6.98]$$

These results immediately follow from the Wald identities (see, e.g. [JAN 07, p. 304]) explaining that:

$$\begin{aligned} E\left[\sum_{n=1}^{N(t)} Y_n\right] &= E[N(t)]E[Y], \\ \operatorname{var}\left[\sum_{n=1}^{N(t)} Y_n\right] &= E[N(t)]\operatorname{var}[Y] + \operatorname{var}[N(t)]\big(E[Y]\big)^2, \end{aligned} \qquad [6.99]$$

where

$$E[N(t)] = \operatorname{var}[N(t)] = \lambda t. \qquad [6.100]$$

Moreover, the process X is a martingale with the minimal filtration iff:

$$\mu + \lambda E[Y] = 0. \qquad [6.101]$$

Indeed, we have:

$$E\left[X(t)|\Im_s\right] = X(s) + E\left[X(t) - X(s)|\Im_s\right]. \qquad [6.102]$$

Moreover, the X process has independent increments, and so:

$$\begin{aligned} E\left[X(t)|\Im_s\right] &= X(s) + E\left[X(t) - X(s)|\Im_s\right] \\ &= X(s) + E[X(t) - X(s)] \end{aligned}. \qquad [6.103]$$

It follows that the process X is a martingale iff $E[X(t)-X(s)]=0$, for all s,t. This last relation is well satisfied as X is a Lévy process.

There exists a similar proposition for the geometric Brownian process.

PROPOSITION 6.4.– If S is a mixed geometric Brownian–Poisson process represented by relation [6.95], then:

$$E(S(t)) = S_0 E^{(\mu+\lambda E[Y])t},$$
$$\operatorname{var}(S(t)) = (E[S(t)])^2 (e^{(\sigma^2+\lambda E[Y^2])t} - 1). \qquad [6.104]$$

Moreover, the process S is a martingale with respect to the minimal filtration iff:

$$\mu + \lambda E[Y] = 0. \qquad [6.105]$$

Similarly, the process $(e^{-rt} S(t))$ is a martingale iff:

$$\mu + \lambda E[Y] = r. \qquad [6.106]$$

6.11.2. *Merton model with jumps*

6.11.2.1. *Definition*

In 1976, Merton [MER 76] extended the Black and Scholes model given in 1973 with the introduction of possible jumps, representing sudden variations of the asset value. In fact, this model is a particular case of a Lévy process introduced in finance 20 years later and motivated some authors [DIB 97] to introduce semi-Markov processes in finance.

Actually, the Merton model only adds a Poisson component to the Black and Scholes model and gives the possibility to have sample paths as given in Figure 6.3.

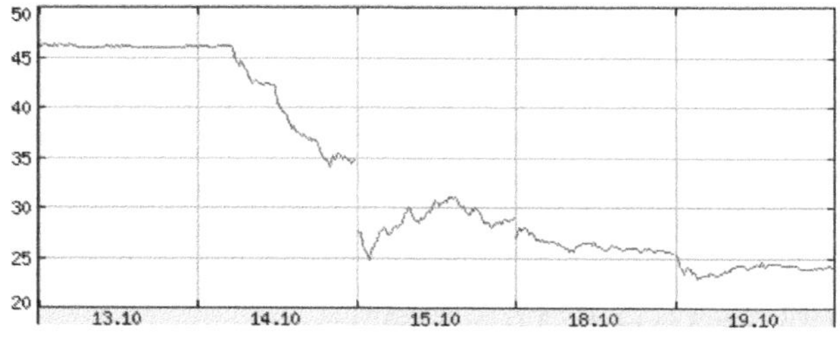

Figure 6.3. *Evolution (end of 2004) of the title of Mercer, stock (source: http://www.yahoo.fr.)*

Frédéric Planchet (ISFA – Université Lyon 1) and Pierre Thérond (Joël Winter & Associés)

On the complete filtered probability space $(\Omega, \Im, (\Im_t), P)$, let the stochastic process of asset prices $S = (S(t), 0 \le t \le T)$ be defined by:

$$\frac{S(t)}{S(0)} = e^{\left(\mu - \frac{\sigma^2}{2}\right)t + \sigma B(t) + \sum_{n=1}^{N(t)} U_k}, \qquad [6.107]$$

where:

– $B = (B(t), t \ge 0)$ is an SBM adapted to the filtration;

– $N = (N(t), t \ge 0)$ is the counting Poisson process of parameter λ and adapted to the filtration such that for all $t > s$, $N(t) - N(s)$ is independent of σ -algebra $\Im_s$; it represents the time occurrence of jumps;

– the r.v. $(U_k, k \ge 1)$ are i.i.d., $N(0, \sigma_u^2)$ adapted to the filtration and represent the amplitudes of the jumps;

– the processes B, N and $U = (U_k, k \ge 1)$ are independent.

REMARK 6.7.– There is a difference in the definition of geometric Brownian motion because in the Merton model the distribution of the jumps is a normal distribution, which can give negative values of the stock with very high negative jumps.

Thus, we have to correct the model to define the jumps by lognormal $U_n = 1 + Y_n$ such that:

$$\ln(1 + Y_n) \prec N\left(\ln(1 + m) - \frac{\alpha^2}{2}, \alpha^2\right).$$

with [6.108]

$$m > -1$$

From the results on the lognormal distribution, it follows that:

$$\begin{aligned} &E[U_n] = E[1 + Y_n] = e^{\ln(1+m)} = 1 + m \Rightarrow E[Y_n] = m, \\ &\operatorname{var}[U_n] = \operatorname{var}[Y_n] = e^{2\ln(1+m)}(e^{\alpha^2} - 1) = (1 + m)^2 (e^{\alpha^2} - 1) \end{aligned}. \qquad [6.109]$$

Henceforth, when we speak about the Merton model, we will always include this correction.

6.11.2.2. *Distribution of S(t)*

Since

$$P\left(\ln\frac{S(t)}{S(0)}\leq x\right)=P\left(\left[\left(\mu-\frac{\sigma^2}{2}\right)t+\sigma B(t)+\sum_{n=1}^{N(t)}U_k\right]\leq \ln x\right), x>0,$$

using the independence assumption between the processes U and B, we get:

$$\begin{aligned}P(\ln\frac{S(t)}{S(0)}\leq x)&=\sum_{n=0}^{\infty}\left(P[\sigma, B(t)+\sum_{k=1}^{n}\ln(1+Y_k)\leq \ln x-\left(\mu-\frac{\sigma^2}{2}\right)t, N(t)=n\right)\\&=\sum_{n=0}^{\infty}\left(P[\sigma B(t)+\sum_{k=1}^{n}(\ln(1+Y_k)\leq \ln x-\left(\mu-\frac{\sigma^2}{2}\right)t|N(t)=n)P(|N(t)=n)\right)\end{aligned} \quad [6.110]$$

From the independence assumptions, we know that:

$$\sigma B(t)+\sum_{n=1}^{n}\ln(1+Y_k)\prec N\left[n\left(\ln(1+m)-\frac{\alpha^2}{2}\right), n\alpha^2+t\sigma^2\right], \quad [6.111]$$

and so:

$$P(\ln\frac{S(t)}{S(0)}\leq x|N(t)=n)=\Phi\left(\frac{\ln x-\left(\mu-\frac{\sigma^2}{2}\right)t-n\left[(\ln(1+m)-\frac{\alpha^2}{2})\right]}{\sqrt{\alpha^2 n+t\sigma^2}}\right). \quad [6.112]$$

Therefore, we finally obtain:

$$P(\ln\frac{S(t)}{S(0)}\leq x)=e^{-\lambda t}\sum_{n=1}^{\infty}\frac{(\lambda t)^n}{n!}\Phi\left(\frac{\ln x-\left(\mu-\frac{\sigma^2}{2}\right)t-n\left[(\ln(1+m)-\frac{\alpha^2}{2})\right]}{\sqrt{\alpha^2 n+t\sigma^2}}\right). \quad [6.113]$$

REMARK 6.8.– From remark 6.7, we know that in the initial Merton model, the jumps have a normal distribution $N(0,\sigma_u^2)$. In this case, the preceding results assume the following form:

$$P\left(\ln\frac{S(t)}{S(0)}\le x\middle|N(t)=n\right)=\Phi\left(\frac{\ln x-\left(\mu-\frac{\sigma^2}{2}\right)t}{\sqrt{\sigma_u^2 n+t\sigma^2}}\right),$$

$$P\left(\ln\frac{S(t)}{S(0)}\le x\right)=e^{-\lambda t}\sum_{n=1}^{\infty}\frac{(\lambda t)^n}{n!}\Phi\left(\frac{\ln x-(\mu-\frac{\sigma^2}{2})t}{\sqrt{\sigma_u^2 n+t\sigma^2}}\right). \qquad [6.114]$$

Moreover, it is possible to prove (see [PLA 05]) that:

$$E\left[\left(\frac{S(t)}{S_0}\right)^p\right]=\exp\left[p\left(\mu-\frac{\sigma^2}{2}\right)t+\frac{p^2\sigma^2}{2}t+\lambda t\left(e^{\frac{p^2\sigma_u^2}{2}}-1\right)\right]. \qquad [6.115]$$

6.11.3. *Stochastic differential equation (SDE) for mixed arithmetic Brownian–Poisson and geometric Brownian–Poisson processes*

It is important to know the SDE governing the mixed arithmetic Brownian–Poisson and geometric Brownian–Poisson processes. To obtain this SDE, we must differentiate [6.94] and [6.96]. However, to do this, it is necessary to recall the definition of the stochastic integral with respect to a Poisson process and to rewrite the classical Itô formula discussed in Chapter 5 and in section 6.6.3. Hereby, we will give the main results (see [ELK 01] for more details):

1) The stochastic integral with respect to a Poisson process of a left continuous function g is defined as follows:

$$\int_s^t g(u)dN(u)=\sum_{i=1}^{N(t)}g(T_i)1_{s<T_i\le t}. \qquad [6.116]$$

2) For a continuous function f of the two variables (t, x) defined on $\mathbb{R}^+\times\mathbb{R}$ having a continuous first partial derivative with respect to x, we have the following Itô formula:

$$f(t,Z(t))-f(0,x)=\int_0^t\frac{\partial f}{\partial x}(u,Z(u-))du$$
$$+\int_{[0,t]}\left[f(u,Z(u-)+Y(u))-f(u,Z(u-))\right]dN(u)$$

where [6.117]

$$Z(t)=x+\sum_{i=1}^{N(t)}Y_i, t\ge 0$$

is the considered compound Poisson process and x a constant.

3) If X is a mixed arithmetic Brownian–Poisson process represented by relation [6.97], then we can apply the result [6.117], but between jumps, we use the classical Itô formula for a Brownian motion with drift to obtain:

$$f(t,X(t))-f(0,x)=$$
$$\int_0^t\left[\frac{\partial f}{\partial t}(s,X(s-))+\mu\frac{\partial f}{\partial x}(s,X(s-))+\frac{1}{2}\sigma^2\frac{\partial^2 f}{\partial x^2}(s,X(s-))\right]ds \quad [6.118]$$
$$+\sigma\int_0^t\frac{\partial f}{\partial x}(s,X(s-))dB(s)+\int_{[0,t]}\left[f(s,X(s-)+Y(s))-f(s,X(s-))\right]dN(s-).$$

In terms of differentiation, we can also write:

$$df(t,X(t))=\left[\frac{\partial f}{\partial t}(t,X(t-))+\mu\frac{\partial f}{\partial x}(t,X(t-))+\frac{1}{2}\sigma^2\frac{\partial^2 f}{\partial x^2}(t,X(t-))\right]$$
$$+\sigma\frac{\partial f}{\partial x}(t,X(t-))dB(t)+\left[f(t,X(t-)+Y(t))-f(t,X(t-))\right]dN(t-). \quad [6.119]$$

EXAMPLE 6.2.– Taking the special case $f(t,x)=e^x$ and with the mixed arithmetic Brownian motion:

$$X(t)=x_0+\mu t+\sigma B(t)+\sum_{n=1}^{N(t)}Y_n \quad [6.120]$$

Formulas [6.119] and [6.119] become:

$$e^{X(t)}-e^x=$$
$$\int_0^t\left[\left(\mu+\frac{\sigma^2}{2}\right)e^{X(s-)}\right]ds$$
$$+\sigma\int_0^t e^{X(s-)}dB(s)+\int_{[0,t]}\left[e^{X(s-)+Y(s))}-e^{X(s-)}\right]dN(s-) \quad [6.121]$$

and

$$de^{X(t)}=\left(\mu+\frac{\sigma^2}{2}\right)e^{X(t-)}dt+\sigma e^{X(t-)}dB(t)+\left[e^{X(t-)+Y(t))}-e^{X(t-)}\right]dN(t-).$$

EXAMPLE 6.3.– Let us consider again that the mixed *geometric Brownian–Poisson process* S is given by:

$$S(t) = S_0 e^{(\mu - \frac{\sigma^2}{2})t + \sigma B(t) + \sum_{n=1}^{N(t)} \ln(1+Y_n)}$$

or [6.122]

$$S(t) = S_0 e^{(\mu - \frac{\sigma^2}{2})t + \sigma B(t)} \prod_{n=1}^{N(t)} (1+Y_n), 0 \le t \le T.$$

Here, the process B is an SBM and the process

$$\left(\sum_{n=1}^{N(t)} \ln(1+Y_n), t \ge 0 \right) \quad [6.123]$$

is a compound Poisson process with parameter λ, the r.v. Y_n having F as the common distribution function with $Y_n > -1$ for all n. Moreover, the two processes are independent.

Since

$$S_t = S_0 e^{X_t} ,$$

we obtain:

$$\frac{S(t)}{S_0} = e^{\left(\mu - \frac{\sigma^2}{2}\right)t + \sigma B(t) + \sum_{n=1}^{N(t)} \ln(1+Y_n)}. \quad [6.124]$$

Therefore, from the results of example 6.2, we successively have:

$$S(t) - S_0 = \mu \int_0^t S(s-)ds + \sigma \int_0^t S(s-)dB(s) + \int_{[0,t]} S(s-)dN(s) \quad [6.125]$$

or under differentiation form:

$$dS(t) = \mu S(t-)dt + \sigma S(t-)dB(t) + Y(t)dN(t),$$
$$S(0) = S_0$$

Of course, for $\lambda = 0$, the jump component disappears, and the last relation gives the classical Black and Scholes model.

6.11.4. *Value of a European call for the lognormal Merton model*

To obtain the call value of maturity T and exercise price K, it suffices to compute:

$$\begin{aligned} C(S,0) &= E_Q\left[\max\{S-K,0\}\right] \\ &= \int_{\mathbb{R}} \{S-K,0\} d_Q P(S(T) \leq S), \end{aligned} \qquad [6.126]$$

where the measure Q corresponds to $\mu = r - \lambda m$.

As

$$P(\ln\frac{S(t)}{S(0)} \leq x | N(t)=n) = \Phi\left(\frac{\ln x - \left(\mu - \frac{\sigma^2}{2}\right)t - n\left[(\ln(1+m) - \frac{\alpha^2}{2})\right]}{\sqrt{\alpha^2 n + t\sigma^2}}\right) \qquad [6.127]$$

given that $N(t) = n$, we can compute the call value using the classical Black and Scholes result for the call, denoted $C(0, S_0, T, K, \sigma^2)$, given here:

$$C\left(0, S_0(1+m)^n e^{-\lambda mT}, T, K, \sigma^2 + n\frac{\alpha^2}{T}\right) \qquad [6.128]$$

and so:

$$C_M(0,S_0) = e^{-\lambda T}\sum_{n=0}^{\infty}\frac{(\lambda T)^n}{n!}C\left(0, S_0(1+m)^n e^{-\lambda mT}, T, K, \sigma^2 + n\frac{\alpha^2}{T}\right) \qquad [6.129]$$

6.12. Complete and incomplete markets

In section 6.2, we have already defined the concept of a *complete market* in stochastic finance as an efficient market in which every contingent claim can be hedged by a self-financing dynamic portfolio investing partly in a bank with a non-risky asset and in the underlying stock. This means that at every time $t, 0 \leq t \leq T$, the portfolio value matches the contingent claim value. In case of a complete market, there exists an equivalent measure Q to P such that the price of the considered contingent claim G of maturity T is given by its present value at time t:

$$E_Q\left[e^{-r(T-t)}G(S(T),T)|\Im_t\right].$$

This result has already been used, for example, in Chapter 5 in the Black and Scholes complete market, replacing the resolution of the Black and Scholes partial differential equation (PDE) by the computation of an integral, provided that we know the distribution of $S(T)$.

We have seen that the Black and Scholes model was always affected by the consequence of the absence of arbitrage (AOA) principle, postulating that without initial capital it is not possible to make a profit at time T without risk, also called *no free lunch with vanishing risk*. Otherwise, the market will not be in equilibrium as the research of more and more profit will be the optimal criterion.

This is why the AOA principle is the basis of almost all models in stochastic finance; nevertheless, in the period of crises, it is clear that intense speculation will destroy this basic assumption and the only possible model for pricing contingent claims is to use the physical probability measure P.

With the AOA assumption, the existence of a martingale measure can be proved, but the unicity of this martingale measure is not always guaranteed, in particular, for incomplete markets [MIY 12].

Schoutens [SCH 03] showed that Lévy market models, other than the Poisson and of course the Brownian cases, are incomplete. As we have seen in the preceding section, this means that there exist a lot of different equivalent martingale measures and, therefore, a lot of different ways to price contingent claims. Therefore, as we have seen in section 6.8.5, it is necessary to add supplementary conditions (see, e.g. [MIY 12]) to have a unique price such as a technical condition like the *Esscher condition* or the *mean-correcting martingale* measure proposed by Schoutens [SCH 03].

Finally, let us point out that in case of a Lévy model without a Brownian component, the price of the considered contingent claim $V(S, t)$ is supposed to be of class C^1 in t and C^2 in S; then, with a risk-neutral martingale measure Q, V satisfies the following partial differential integral equation (PDIE), extending the classical Black and Scholes PDE:

$$rG(S,t) = \gamma \frac{\partial}{\partial S} G(S,t) + \frac{\partial}{\partial t} G(S,t) + \int_{-\infty}^{+\infty} \left[G(S+y,t) - G(y,t) - y \frac{\partial}{\partial S} G(S,t) \right] v_Q(dy) \text{ with}$$

the boundary condition of type $G(S,T) = F(S)$.

6.13. Conclusion

If the approach of option pricing using Lévy processes really extends the classical Black and Scholes approach, but, except with the Merton model, we lose the unicity of the risk-neutral measure. This leads to other problems, which must still be studied, and simulation is the only way to use such models from a practical point of view but that is not very useful for traders working on real time in market rooms. Complementary results can be found in [SCH 03] and [PAS 11].

Finally, let us point out again that the Lévy processes still have independent increments, an assumption that is also criticized from the viewpoint of realistic stochastic models in finance. Moreover, we have seen that Lévy processes are time and space homogeneous continuous-time Markov processes, and so they cannot be compared to semi-Markov models. Indeed, semi-Markov models have not independent increments and, moreover, they can be used in a non-homogeneous time version which can be very useful for approaching real-life problems with such semi-Markov models.

7

Actuarial Evaluation, VaR and Stochastic Interest Rate Models

In the framework of international recommendations of Solvency II, there are strong modifications in the risk management for insurance companies involving techniques issuing from Basel II and III, such as value-at-risk (VaR) techniques, and also stronger approaches of the own risk of the considered company with own solvency risk assessment (ORSA), thereby changing the concept of actuarial evaluation. For this, stochastic models of interest rate evolution can be used. We present here some main ideas of this subject and more details can be seen in [JAN 09b, COR 12] and [GEH 15, DEV 12].

7.1. VaR technique

The value-at-risk (VaR) technique was developed by J.P. Morgan and Company in 1994 in the follow-up of Basel I prudential rules related to the quantification of credit and market risks. It was distributed under the name of Riskmetrics as a way to measure the protection against the shortfall risk, that is the critical risk of not having enough equity against facing a bad situation.

The aim of the VaR theory is to find, for a given risk represented by the non-negative r.v. X, an amount of equity such that the probability of having a loss larger than this value is very small, for example 1 or 0.5% and thus compatible with the attitude of the management against risk.

It is clear that the calculation of VaR values depends on the considered risk and particularly on its distribution function.

Of course, this determination always depends on the time horizon on which we are working: 1 day, 1 week, 1 month or 1 year for Solvency II or even more for ORSA studies.

For a risk X having a general d.f. F_X on the considered time horizon T, the $VaR_\alpha^T(X)$ or simply VaR_α at level α satisfies the following equality:

$$F_X(VaR_\alpha) = \alpha \qquad [7.1]$$

and if the function F_X is strictly increasing, we obtain:

$$F_X^{-1}(\alpha) = VaR_\alpha. \qquad [7.2]$$

So, from a statistical point of view, the indicator $VaR_\alpha(X)$ is nothing else but the quartile at level (1-α) of the distribution function of the risk X.

Relation [7.2] gives the way to calculate the VaR value at a given level provided that we know the d.f. of X. Unfortunately, the knowledge of this function is in general not easy, and therefore a number of estimation methods have been constructed to circumvent this problem.

EXAMPLE 7.1.–

If the r.v. X has a normal distribution with mean m_X and variance σ_X^2, the VaR value at level α is given by the quartile of order $(1-\alpha)$ (with α small) of the d.f. of X. As $P(X \leq z_\alpha \sigma_X + m_X) = \alpha,$ where z_α satisfies $\Phi(z_\alpha) = 1-\alpha$ and Φ represents the d.f. of a reduced normal distribution, we have that:

$$VaR_\alpha = z_\alpha \sigma_X + m_X. \qquad [7.3]$$

Let us suppose that a financial institution has in its portfolio shares with an individual value of €700.

On the basis of historical data, the global return on a time period of 1 year for one share of value S, for example, is estimated as having the following normal distribution:

$$S(1) - S_0 \prec N(60,1600).$$

It follows that the loss on the period has a normal distribution of mean –60 and standard deviation 40.

Using result [7.3], we obtain the VaR values given in Table 7.1 with different security or probability levels.

A	0.95	0.99	0.999	0.9999
VaR	5.796	33.052	64	88

Table 7.1. *VaR values for one asset with the normal distribution*

The interpretation of these results uses the frequency interpretation of probability stating that the probability of an event can be seen as the ratio of the "favorable" cases, i.e. the realization of the considered event, over the total number of realizations, this last realization assumed to be large so that this interpretation is confirmed by the law of large numbers.

So, with a level of 0.999, after 1 year there is one chance in 1,000 that the observed loss is over €64 per share.

If level 0.9999 is imposed, with one chance in 10,000, the loss per action is larger than €88, which is 40% larger than with the preceding level.

EXAMPLE 7.2.– VaR for the truncated normal case

With definition [7.1] of the VaR, the loss may be negative, in which case it is a gain. This is correct except that, for example, if we use the VaR concept for a share bought at amount S_0, of course, the loss has S_0 as maximum value which corresponds to the assumption of non-negativity of the r.v. X represented the considered risk.

Also, it may be better to use the truncated normal distribution for the loss:

$$S_0 - S(T) \prec N_{tr(S_0)}(m_T, \sigma_T^2)$$

so, that the loss is strictly less than S_0 with probability 1.

It follows that the distribution function of the loss is given by:

$$\frac{\Phi(\frac{x-m_T}{\sigma_T})}{\Phi(\frac{S_0-m_T}{\sigma_T})}, x < S_0. \tag{7.4}$$

Consequently, relation [7.3] becomes:

$$\frac{\Phi(\frac{Var_{tr,\alpha}-m_T}{\sigma_T})}{\Phi(\frac{S_0-m_T}{\sigma_T})} = \alpha \tag{7.5}$$

and so

$$\frac{Var_{tr,\alpha}-m_T}{\sigma_T} = z_{\alpha\Phi\left(\frac{S_0-m_T}{\sigma_T}\right)}$$

or [7.6]

$$Var_{tr,\alpha} = m_T + \sigma_T z_{\alpha\Phi\left(\frac{S_0-m_T}{\sigma_T}\right)}.$$

EXAMPLE 7.3.– Numerical example

Let us return to numerical example 7.1 for which we assume that $S_0 - S(1) \prec N(-60,1600)$.

Now, let us use the truncated normal distribution and apply result [7.6]. Table 7.2 presents the following results.

Alpha	0.95	0.99	0.999	0.9999
VaR	5.94	33.059	63.609	88.76

Table 7.2. *VaR values for one asset with the truncated normal distribution*

We see that these results are very close to the results given in Table 7.1, obtaining without truncating the normal distribution. This is due to the fact that the mean is –60 and thus both distributions are more or less identical, but the situation may be different for means close to zero.

EXAMPLE 7.4.–

Let us now consider an example of an insurance company, which has a portfolio of a protection business. This portfolio is assumed to have a loss ratio having a normal distribution of mean 65 and standard deviation 40. Then, VaR values over 1-year horizon at different security of probability levels are given in Table 7.3.

A	0.95	0.99	0.999	0.9999
VaR	0.794	28.054	58.609	83.60

Table 7.3. *VaR values for a protection business portfolio with the normal distribution*

With a level of 0.999, after 1 year there is one chance in 1,000 that the observed loss is over 58,609.

If level 0.9999 is imposed, with one chance in 10,000, the loss per action is larger than 83,760.

7.2. Conditional VaR value

To give a better risk indicator, we can also introduce the *conditional VaR value* (CVaR) indicator defined as:

$$E(X - VaR_\alpha(X)) | X > VaR_\alpha(X)) \qquad [7.7]$$

representing the excess of the risk if the observed claim is larger than the VaR.

For the CVaR, we have to use the following general result:

$$E(X-u)|X>u) = \frac{\int_u^\infty (1-F_X(x))dx}{1-F_X(u)} \qquad [7.8]$$

EXAMPLE 7.5.– Explicit forms VaR and CVaR in Pareto case

From

$$F(x)=\begin{cases}1-\left(\dfrac{\theta}{x+\theta}\right)^{\beta}, x\geq 0,\\ 0, x<0,\end{cases}$$

we obtain:

$$1-\left(\frac{\theta}{VaR_{\alpha}(X)+\theta}\right)^{\beta}=\alpha$$

or [7.9]

$$(1-\alpha)^{1/\beta}=\frac{\theta}{VaR_{\alpha}(X)+\theta}$$

and finally:

$$VaR_{\alpha}(X)=\frac{\theta}{(1-\alpha)^{1/\beta}}-\theta.$$

For $\theta=\beta=1$, this value becomes:

$$VaR_{\alpha}(X)=\frac{\alpha}{1-\alpha}. \qquad [7.10]$$

And for $\alpha=0{,}99$:

$$VaR_{\alpha}(X)=\frac{0{,}99}{0{,}01}=99 \qquad [7.11]$$

For the CVaR, we have to use the following general result:

$$E(X-u)|X>u)=\frac{\int_{u}^{\infty}(1-F_X(x))dx}{1-F_X(u)} \qquad [7.12]$$

where for the Pareto case, the function F_X is given by the first result of relation [7.9]. This gives the following result:

$$CVaR_{\alpha}(X)=\frac{\theta+VaR_{\alpha}(X)}{\beta-1}, \qquad [7.13]$$

result valuable only if $\beta>1$.

From the last relation of result [7.9], we obtain:

$$\begin{aligned} CVaR_\alpha(X) &= \frac{\theta + \dfrac{\theta}{(1-\alpha)^{1/\beta}}}{\beta - 1} \\ &= \theta \frac{(1-\alpha)^{1/\beta} + 1}{(1-\alpha)^{1/\beta}(\beta - 1)}. \end{aligned} \quad [7.14]$$

EXAMPLE 7.6.– VaR for extreme value distributions

Using the general form of extreme value distributions:

$$\begin{aligned} &a(y) = \exp\left[-(1-\tau y)^{1/\tau}\right], \\ &\left(\tau = -\frac{1}{\alpha} \text{ for Fréchet, } \tau = \frac{1}{\beta} \text{ for Weibull and } \tau = 0 \text{ for Gumbel}\right), \end{aligned} \quad [7.15]$$

we have:

$$\exp\left[-(1-\tau VaR_\alpha)^{1/\tau}\right] = \alpha$$

and so

$$\frac{1-(-\ln\alpha)^\tau}{\tau} = VaR_\alpha. \quad [7.16]$$

So for the Fréchet case for which:

$$x = 1 - \tau y, \beta = -\frac{1}{\tau}$$

we obtain:

$$\begin{aligned} &x = 1 - \tau y, \beta = -\frac{1}{\tau} \\ &VaR_\alpha(F) = 1 - (1 - (-\ln\alpha)^{-\frac{1}{\beta}}) \end{aligned} \quad [7.17]$$

and

$$VaR_\alpha(F) = (-\ln\alpha)^{-\frac{1}{\beta}}.$$

And for $\alpha = 0.99$:

$$VaR_{\alpha}(X) = \frac{0.99}{0.01} = 99 \qquad [7.18]$$

EXAMPLE 7.7.–

We consider a Pareto distribution with $\theta = 1$ and different values for β.

Table 7.4 gives the VaR values using the last result of [7.9].

Beta \ Alpha	0.95	0.99	0.995	0.9995
1.1	14.2319162	64.7933225	122.550768	1001.15837
1.2	11.1392446	45.4158883	81.7037108	562.453823
1.3	9.01826013	33.510729	57.8874941	345.141637
1.5	6.368063	20.5443469	33.1995189	157.740105
1.1	14.2319162	64.7933225	122. 50768	1001.15837
1.2	11.1392446	45.4158883	81.7037108	562.453823
1.3	9.01826013	33.5510729	57.8874941	345.141637
1.5	6.368063	20.5443469	33. 995189	157.740105

Table 7.4. *VaR values, for example*

Table 7.5 gives the corresponding VaR values with a Gaussian model when it is possible.

β	Mean	Variance	0.95	0.99	0.995	0.9995
1.1	10					
1.2	5					
1.3	3.333333					
1.5	2					
2	1					
2.5	0.666666	3.333333	3.503078	4.913977	5.369466	6.674319
3	0.5	1.5	2.514526	3.349182	3.654733	4.530055
5	0. 25	0.416666	1.311748	1.75165	1.912690	4.530055

Table 7.5. *Corresponding VaR values with a Gaussian model*

Finally, Table 7.6 gives the ratio VaR – mean for α = 0.995.

Alpha / Beta	0.995	Mean	VaR-mean	Ratio
1,1	122.550768	10	112.550768	11.2550768
1,2	81.7037108	5	76.7037108	15.3407422
1,3	57.8874941	3.33333333	54.5541607	16.3662482
1,5	33.1995189	2	31.1995189	15.5997595
2	13.1421356	1	12.1421356	12.1421356
2,5	7.32553207	0.66666667	6.65886541	9.98829811
3	4.84803548	0.5	4.34803548	8.69607095
5	1.88539981	0.25	1.63539981	6.54159925

Table 7.6. *Ratio VaR – mean for α=0.995*

With result [7.14], Table 7.7 gives the CVaR values.

Alpha / Beta	0.95	0.99	0.995	0.9995
1.1	152.319162	657.933225	1235.50768	10021.5837
1.2	60.6962231	232.079442	413.518554	2817.26911
1.3	33.3942004	115.170243	196.291647	1153.80546
1.5	14.736126	43.0886938	68.3990379	317.48021
2	4.47213595	10	14.1421356	44.7213595
2.5	2.20963601	4.2063823	5. 5035472	13.9418607
3	1.35720881	2.32079442	2.92401774	6.29960525
5	0.45514105	0.62797161	0.72134995	1.14326263

Table 7.7. *Some CVaR values*

REMARK 7.1.– The use of the VaR for insurance companies

As shown below in section 7.3, Solvency II is based, such as Basel agreements, on a three-pillar approach:

– pillar I consists of the quantitative requirements (for example, the amount of capital an insurer should hold);

– pillar II sets out requirements for the governance and risk management of insurers, as well as for the effective supervision of insurers;

– pillar III focuses on disclosure and transparency requirements.

Insurance companies use the VaR in the framework of the Pillar I of Solvency II in order to calculate the regulatory capital. The parameters to be used are $\alpha = 0.995$ and T=one year. In this case, the confidence level and the time horizon are defined by the regulator, are common to all insurance companies. This allows for consistency and comparability between insurers.

Insurance companies also calculate their risk profile in the framework of the Pillar II of Solvency II. To define their own risk and solvency assessment (ORSA), they must choose a tool to measure risk (VaR, CVaR, etc.), a confidence level and a time horizon to better quantify their risk profile.

7.3. Solvency II

Solvency II introduces new solvency requirements, which are more risk-sensitive and sophisticated than Solvency I. In the framework of Solvency II, business is considered in runoff.

The main indicators can be computed either with a *standard approach* fully described in the Solvency II rules or with *partial* or *total internal models*.

The capital requirement is determined in a two-level approach already presented in Chapter 1.They correspond to VaR with different levels of confidence on a 1-year horizon. The first level is the *minimum capital requirement* (MCR), which is the capital representing the threshold that triggers ultimate supervisory measures in the event that it is breached. The second level is the *solvency capital requirement* (SCR), which can be calculated using the standard formula, a full internal model or a partial internal model. These two levels of capital requirements are described in the following sections.

7.3.1. *The SCR indicator*

To calculate the CSR (see [COR 12, GEH 15]) insurance companies must use the risk classification of Solvency II, which includes risk modules (market, health, default, life, non-life, (intangible)) and risk submodules (such as mortality, longevity, disability/morbidity, lapse, expense, revision and CAT for the life module). This calibration is maintained at the module level by stress tests or closed formulas.

Figure 7.1 shows the Solvency II risk cartography, established for the Quantitative Impact Study no. 5 (QIS5).

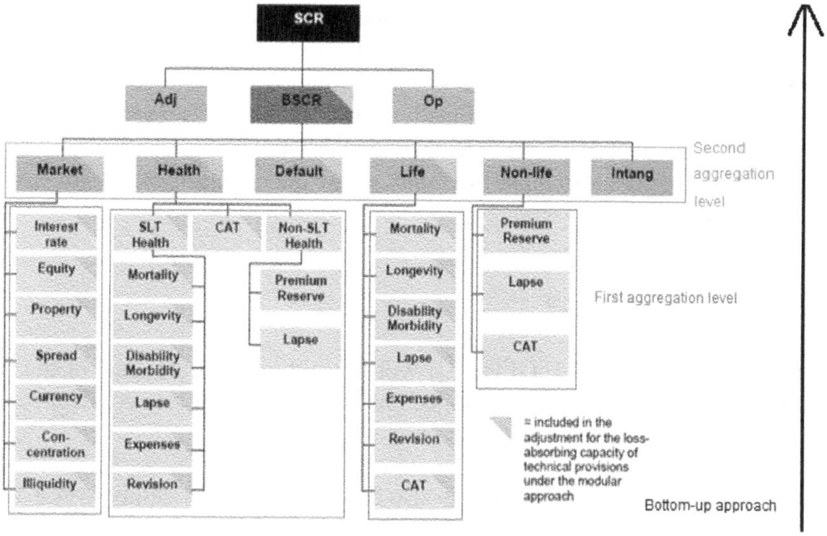

Figure 7.1. *Risk classification (source: QIS5 technical specifications)*

In fact, capital charges are determined using a bottom-up approach. First, the capital requirements are evaluated for each risk submodule. Second, the capital requirements of submodules are aggregated using a prescribed correlation matrix to obtain the capital requirement of risk modules. Third, another aggregation using a prescribed correlation matrix is done between the capital requirements of risk modules to give the basic SCR (BSCR). Finally, a separate loading for operational risk is added to obtain the company SCR.

When the company uses an *internal model*, it has to create a stochastic model with structures and relationships that best depict the company's business. The description is done for the whole structure of the company in a *full internal model* and for some parts of this structure in partial internal models. The risks incorporated into an internal model are likely to be the same as those included in the standard model. The calculation of the SCR using an internal model is based on the distribution of equities in a year, as described in [DEV 09]. To calculate the SCR, two steps are needed:

– to project assets and liabilities under the historical probability in order to evaluate the quantile 99.5% of the net asset value at time t = 1;

– to discount at time t = 0 the value of the quantile 99.5% in order to calculate the amount of capital, which invested at risk-free rate, will enable the insurance company to avoid bankruptcy at time t = 1 in 99.5% of cases.

Internal models enable the insurance company to quantify risk and determine the capital requirement on the basis of the company's specific risk. Nevertheless, they require expertise and resources for model building, calibration, validation, interpretation and communication. Finally, they require supervisory approval but clearly the ultimo goal is to provide SCR and MCR values lower than with the standard approach, giving so more flexibility to use the excess of capital for strategic uses of the company.

7.3.2. *Calculation of MCR*

The MCR (see [COR 12, GEH 15]) will be calibrated to the VaR of the basic own funds of an insurance or reinsurance undertaking subject to a confidence level of 85% over a 1-year period.

Thus, the calculation of the MCR is also based on a VaR of equity over a 1-year horizon, but at a different level of confidence.

Let us just add that the MCR has an absolute threshold, which depends on the nature of the business. This threshold should be between 25 and 45% of the SCR (whatever the method of calculating it). More details on this threshold are given in the implementing measures on Solvency II.

To summarize, let us say that the MCR is calculated in three steps:

1) The linear MCR is the sum of the MCR life and MCR non-life, which are in both cases linear formulas. The MCR life depends on technical provisions and capital at risk. The MCR non-life depends on technical provisions and premiums over the last 12 months.

2) The combined MCR compares the linear MCR with 25 and 45% of the SCR.

3) The final MCR is the maximum between the combined MCR and a threshold fixed by the regulator.

With this calculation method, the MCR is obviously dependent on the SCR. The underlying assumption is that 35% of the SCR corresponds to an 85% VaR over 1 year, which can be criticized.

Figure 7.2 sums up the two levels of capital requirements in the framework of Solvency II.

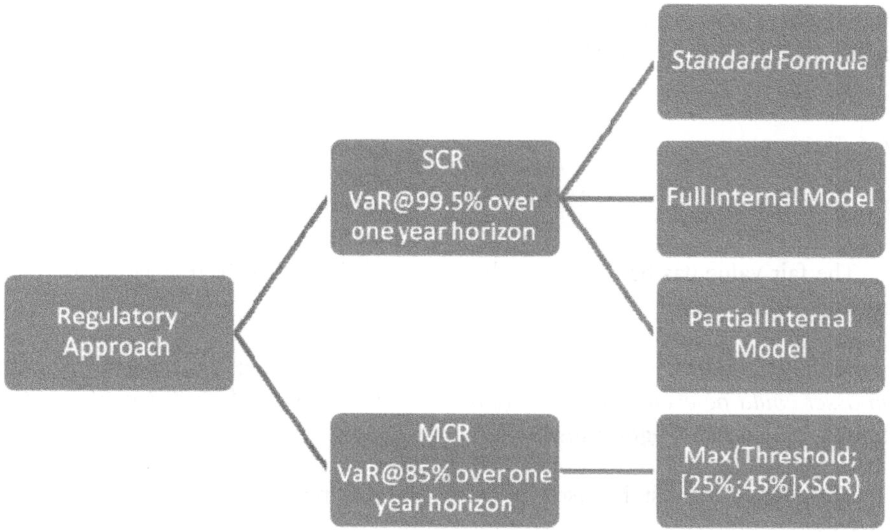

Figure 7.2. *The two levels of capital requirements in the Solvency II framework*

7.3.3. *ORSA approach*

Although Solvency II gives guarantees to the regulatory concerning the financial solidity of the considered insurance company, it is nevertheless a fact that these rules do not take into account the strategy of the company and its main objectives, in particular concerning the considered time horizon of 1 year, though most contracts have medium and long terms. So, the company must be able to guarantee security not only on 1 year, but also for larger time horizons.

Therefore, it is necessary to define a management tool that provides an alternative assessment of the solvency relative to Pillar I.

To do so, insurance companies are encouraged in Solvency II Directive to implement their own risk solvency assessment (ORSA).

To better assess its risk profile, the insurance company has to define the risk cartography that corresponds at best to reality and to choose the risk metric that corresponds at best to its problematic.

The choice of the risk metric is a key element in the ORSA. The insurance company has to choose a risk indicator such as VaR, CVaR and SCR, a time horizon, for example 5 or 10 years or even more, compatible with its management horizon and a level of confidence not necessarily 99.5% but less such as (see [COR 12, GEH 15]).

7.4. Fair value

7.4.1. *Definition*

The fair value has been imposed by the new IFRS norms to replace the obsolete booking value.

The IFRS definition of the fair value is the following one: "*the amount for which an asset could be exchanged, or a liability settled, between knowledgeable, willing parties in an arm's length transition*".

This concept must be used in particular for pension funds with the norm IAS 19.

Also, Solvency II introduces the two concepts: the *mark-to-market* and the *mark-to-model* evaluations.

For the first concept, for an asset available at time t, its fair value is just the value on the stock exchange of this asset provided the market is liquid enough to realize the transaction; the *bid price* is retained for emission of the asset and the *ask price* for its purchase.

Sometimes, it is not easy to compute this market value, for example for computer material or financial assets in an illiquidity market. For this case, we can apply the mark-to-model evaluation consisting of building a method for the evaluation of the asset value using, for example, an evaluation of all the future cash flows generated by this asset at the market rate or using an adapted theoretical model which must be of course agreed by the prudential authorities.

For the need of applying rules of Solvency II or Basel II or III, it is often necessary to predict market values in the future and this is not possible without *stochastic models*.

Also, in the next sections, we will show how this can be done.

7.4.2. *Market value of financial flows*

A *financial flow* $F = \{(F_j, t_j), j = 1,...n\}$ gives n successive amounts of cash F_j to be received or paid at successive times $t_j, j = 1,...,n$ $(t_1 < t_2 < ... < t_n)$.

EXAMPLE 7.8.–

A bond of *nominal value P* with *coupons of value C* and *maturity date s + S* gives the right for the investor buying this bond at time s, to receive the coupon value C at times $\{s+1, s+2, \ldots, s+S\}$ and the nominal value P at time $s + S$. It follows that the successive cash flows of this investment are given by:

– at times s , $s+ 1,...,$ $st + S\text{-}1$: the coupon value C;

– at time $s + S$: the amount $P+C$.

If $A(t,T)$ $(s \le t \le s+S)$ represents at time t the value of the bond issuing at time $t + T$, the main problem is to evaluate its fair value in view of comparing it to the proposed market value at time t.

A *zero-bond*, an investment made, for example at time s and of maturity date $s + S$, is a very simple investment for which it is paid the sum $P(s,S)$ at time s in view of receiving €1 at the maturity date $s + S$.

Thus, we can calculate the value of the above bond $A(t,T)$ with the following formula:

$$A(t,T) = P(t,1)C + P(t,2)C + ... + P(t,T-1)C + P(t,T)(C+P). \qquad [7.19]$$

7.4.3. *Yield curve*

It is well known that the interest rate for a deposit at time t depends not only on this time t, but also on its maturity T, so that this annual rate can be written as $i(t,T)$. These rates are related to a given market, for example the market of bonds and therefore using these data, we can compute the so-called *market value* of a given flow F.

For a fixed time t, the graph of the function $T \mapsto i(t,T)$ represents the *yield curve* at time t.

Figure 7.3 provides a real life example.

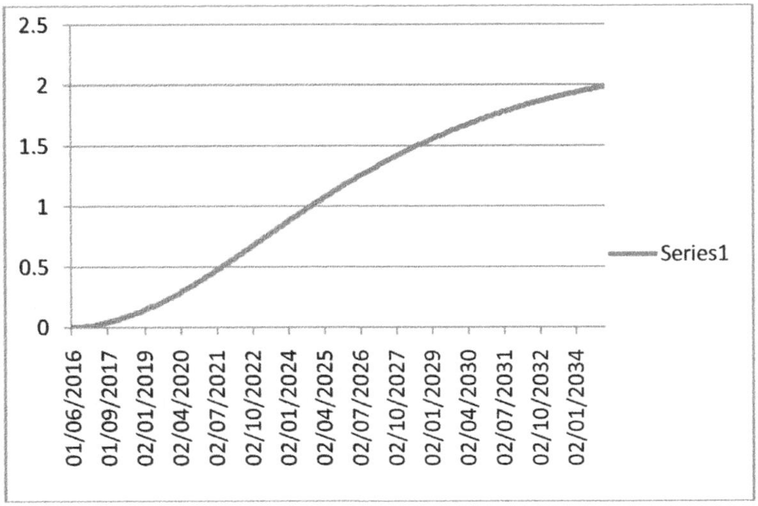

Figure 7.3. *Yield curve in November 2014 in year maturities up to May 2020 (source: IA, France)*

From these data, we see that the spread between short- and long-term interest rates is more or less 2%.

So, given such a yield curve, we obtain the following value for a zero-bond:

$$P(t,T) = (1 + i(t,T))^{-T} \tag{7.20}$$

The computations of the different interest values *i(t,T)* are generally observed in a financial market where we can observe the market values of the zero-bonds. Therefore by using formula [7.19] for different bonds of different maturity times, we can calculate the values of the zero-bonds according to the market values of the observed bonds.

EXAMPLE 7.9.– Let us consider the case of $T = 2$ and let us suppose that we have two bonds, the first with a coupon of 1.2%, with 100 as nominal value and with 1 as maturity, the second with a coupon of 1.4%, with 100 as nominal value and with 2 as maturity. The market values of these two bonds at time T are, respectively, of 100 and 102.

Using formula [7.19] twice, we obtain:

– $100 = P(t,1)(1.2 + 100)$;

– $100 = P(t,1)(1.4) + P(t,2)(1.4 + 100)$.

From the first equation, we obtain $P(t,1)$:

$$P(t,1) = \frac{100}{101.2}(= 0.9888)$$

and then from the second equation, the value of $P(t,2)$:

$$P(t,2) = \frac{100 - 0.9888 \times 1.4}{101.4}(= 0.9923).$$

Consequently, the yield rates for 1 and 2 years are given by:

$$i(t,1) = (0.9888)^{-1} - 1 = 1.132\%,$$

$$i(t,2) = (0.9923)^{-1/2} - 1 = 1.401\%.$$

Of course, in practice, this method needs a bond market liquid enough to have all the data available for all maturities, and moreover a statistical treatment with the least squares method can be used to improve the method.

7.4.4. *Yield to maturity for a financial investment and a bond*

Let us consider a financial investment of present value C generating the following financial flow:

$$F = \{(F_j, t_j)\, j = 1, \ldots, n\}. \qquad [7.21]$$

The *yield to maturity* is the constant discount rate or actuarial rate $i(F)$, solution of the polynomial equation:

$$C = \sum_{j=1}^{n} (1 + i(F))^{-t_j} F_j. \qquad [7.22]$$

Using the traditional Newton interpolation with the nominal or coupon rate as initial value, this solution is easily given.

For the particular case of a bond of *subscription price* A at time t and maturity time $t + T$, and with *coupon value* C and *nominal value* P, the corresponding financial flow is given by:

– at times t+1,...,t+T-1: payment of the coupon C;

– at time t+T: payment of the coupon C and of the nominal value.

Equation [7.19] becomes:

$$A(t,T)=(1+i(F))^{-1}C+(1+i(F))^{-2}C+\dots \\ +(1+i(F))^{-(t+T-1)}C+(1+i(F))^{-(t+T)}(C+P). \qquad [7.23]$$

Once more, let us point out that the yield to maturity $i(F)$ is a function of t and T. This is why we use the notation $i(t,T)$ and in fact the equation $I = i(t,T)$ represents a surface in a three-dimensional (3D) space.

To do forecasts, it is necessary to build models for the evolution of such a surface. The most interesting are of course stochastic models, and in the next section, we briefly recall such most used models.

7.5. Dynamic stochastic time continuous time model for instantaneous interest rate

7.5.1. *Instantaneous deterministic interest rate*

Here, we will use the traditional deterministic continuous time model (DCTM) for an investment on $[0,T]$ of amount $C(t)$ at time t producing a continuous yield of rate $r(s)$ at time s.

So, on the "small" time interval $[s,s+\Delta s]\subset[0,T]$, one monetary unit at time t produces at the end of the interval a yield of value $r(s)\Delta s$. This rate is called the *continuous time instantaneous rate* or, in short, the *instantaneous rate* for an investment at time 0 and of maturity time T.

Let $C(s)$ be the capitalization value of $C(t)$ at time $s, s>0$.

From the definition of the instantaneous rate, it is clear that:

$$C(s+\Delta s)=C(s)+r(s)C(s)\Delta s \qquad [7.24]$$

With traditional limit reasoning, we obtain the following relation:

$$\frac{C'(s)}{C(s)}=r(s) \qquad [7.25]$$

and by integration:

$$C(s)=C(t)e^{\int_t^{t+s} r(u)du}.$$

In particular, at maturity, we obtain:

$$C(T)=C(0)e^{\int_t^{t+T} r(u)du}. \quad [7.26]$$

Conversely, given the value of *C(T)*, we can compute the present value or the market value at time t by the following relation:

$$C(t)=C(T)e^{-\int_t^{t+T} r(u)du}. \quad [7.27]$$

For the particular case of a constant function for r: $r(t)=\delta$, relation [7.26] becomes the well-known result of classical financial algebra:

$$C(t)=C(0)e^{\delta T}. \quad [7.28]$$

7.5.2. *Yield curve associated with a deterministic instantaneous interest rate*

For the preceding section, we know that for an investment of €1 at time *t* and of maturity time *t* + *T*, the capitalization value at maturity is given by:

$$e^{\int_t^{t+T} r(u)du}. \quad [7.29]$$

Using the *yield curve* $T \mapsto i(t,T), T \geq 0$, for a fixed *t*, corresponding to this investment for which *i*(*t*, *T*) represents the corresponding annual interest rate on [*t*,*t*+*T*] given the same capitalization values as [7.29], we obtain:

$$\left(1+i(t,T)\right)^T = e^{\int_t^{t+T} r(u)du} \quad [7.30]$$

and so:

$$i(t,T)=e^{\frac{1}{T}\int_t^{t+T} r(u)du}-1. \quad [7.31]$$

The constant instantaneous rate $\delta(t,T)$ on $[t,t+T]$ corresponding to this yield curve is defined as follows:

$$e^{\int_t^{t+T}\delta(t,T)du} = e^{\int_t^{t+T}r(u)du}$$

or

$$e^{\delta(t,T)T} = e^{\int_t^{t+T}r(u)du}, \quad [7.32]$$

i.e.:

$$\delta(t,T) = \frac{1}{T}\int_t^{t+T} r(u)du.$$

For a constant rate, we obtain:

$$i(t,T) = e^{\frac{1}{T}\int_t^{t+T}\delta du} - 1$$

or $\quad [7.33]$

$$i(t,T) = e^{\delta} - 1.$$

From this last relation, we obtain:

$$\delta = \ln(1+i). \quad [7.34]$$

7.5.3. *Dynamic stochastic continuous time model for instantaneous interest rate*

In finance, it is well known that the future values of the rates are uncertain as there is a large influence from many financial and economic parameters, depending also on worldwide political situation.

It follows that deterministic models are unsatisfactory and therefore the new discipline of mathematical finance called "stochastic finance" was born from the results of [SAM 65] and [BLA 73].

In this section, we will briefly recall the two most important stochastic models used in practice: the *Ornstein–Uhlenbeck–Vasicek* (*OUV*) model and the *Cox, Ingersoll and Ross (CIR)* model presented, for example, in *[DEV 12].*

7.5.4. *The OUV stochastic model*

i) The model

As usual, we consider a complete filtered probability space $(\Omega,\Im,(\Im_t),P)$ on which all the defined stochastic processes will be adapted, in particular, the following discrete OUV model with two parameters a and b starts with the following stochastic dynamic for the spot rate process $r=(r(t),t\geq 0)$:

$$r(t)\approx b+(r_0-b)e^{-at}+\sigma e^{-at}\sum_{i=0}^{n}e^{at_i}\left[B(t_{i+1})-B(t_i)\right],$$

to the following subdivision of the interval $[0,t]$

$(B(t_i), i=0,...,n)$ independent,

$B(t_i)$ normal distributed with mean 0 and variance t_i.

Letting the norm of the considered time subdivision tending to 0, results of Chapter 5 on stochastic calculus show that the process r satisfies the following stochastic differential equation (SDE in short) where B is a standard Brownian motion (see Chapter 5).

$$\begin{aligned}&dr(t)=a(b-r(t))dt+\sigma dB(t),\\&r(0)=r_0.\end{aligned}\qquad [7.35]$$

This means that r is a special diffusion process extending the deterministic OUV case depending on four parameters: a,b,r_0,σ, assumed to be constant and known.

Using Itô's calculus, it is possible to show that the unique solution of this SDE is given by:

$$r(t)=b+(r_0-b)e^{-at}+\sigma e^{-at}\int_0^t e^{as}dB(s).\qquad [7.36]$$

ii) Model and parameter interpretation

From result [7.36] and the traditional rules of Itô differentiation seen in Chapter 5, we obtain the *mean of* $r(t)$:

$$m(t)=E\left[r(t)\right]=b+(r_0-b)e^{-at}.\qquad [7.37]$$

So, this means $m(t)$ tends toward b for $t \to \infty$, result giving the meaning of the first parameter a of this model as the *anticipated spot rate for long term.*

Concerning the *variance* of $r(t)$, we still use Itô differentiation:

$$\operatorname{var} r(t) = \operatorname{var}\left(\sigma e^{-at} \int_0^t e^{as}\, dB(s) \right). \qquad [7.38]$$

So:

$$\operatorname{var} r(t) = \sigma^2 e^{-2at} \int_0^t e^{2as}\, dB(s), \qquad [7.39]$$

and finally:

$$\operatorname{var} r(t) = \frac{\sigma^2}{2a}\left(1 - e^{-2at}\right). \qquad [7.40]$$

This last result shows that the variance of $r(t)$ as t tends to $+\infty$, and moreover this asymptotic variance is a linear function of λ.

So, this variance is smaller for a weakly volatile market σ and larger for a market with large volatility, in conformity with empirical studies.

Parameter a has an opposite effect: large (small) values of a given smaller (larger) values of the variance of r(t).

To conclude, we see that:

1) the variance of $r(t)$ is increasing with time, confirming the fact that the uncertainty on the rate values increases with time;

2) the larger parameter σ, called *volatility*, is, the greater the impact of randomness;

3) the larger parameter a, called *convergence parameter*, is, greater the convergence of the spot rate toward b.

iii) Marginal distribution of r(t), fixed t

To calculate the distribution of $r(t)$, for all fixed t, it suffices from relation via relation [2.40] to calculate one of the r.v. of $X(t)$ defined by:

$$X(t)=\int_0^t e^{as}dB(s). \qquad [7.41]$$

It is possible to prove (see [JAN 09b]) that X(t) has, for each fixed t, a normal distribution and therefore from results [5.30], [6 31] and [5.34], we finally get:

$$r(t)\prec N\left(b+(r_0-b)e^{-at},\frac{\sigma^2}{2a}\left(1-e^{-2at}\right)\right). \qquad [7.42]$$

EXAMPLE 7.10.–

r_0	0,02	b	0,04
A	5		
σ	0,07		
Δt	0,01		

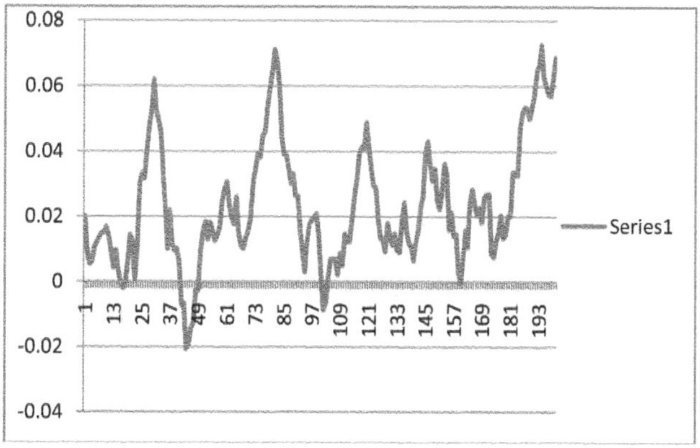

Figure 7.4. *A sample path for an OUV mode*

7.5.5. *The CIR model*

7.5.5.1. *The model*

In 1985, Cox *et al.* [COX 85] presented a new model for the temporal structure of interest rates for which the inconvenience of having negative values, as is the case for the OUV model, was dropped. It is amazing that nowadays, this situation appears.

To obtain the model, the authors introduced a factor $\sqrt{r(t)}$ in the coefficient of dB, and this is why their model is also called the *square root model.*

In discrete time, we have the following model:

$$\left\{\begin{aligned} & r(t_{i+1},\omega) = r(t_i,\omega) + a(b - r(t_i,\omega)(t_{i+1} - t_i) \\ & \qquad + \sigma\sqrt{r(t_i,\omega)}\left[B(t_{i+1},\omega) - B(t_i,\omega)\right], \\ & r(0) = r_0, t_i \in \{t_0,\ldots,t_n\}, t_0 = 0, t_n = T. \end{aligned}\right\}. \tag{7.43}$$

This system being recursive can easily be solved for simulated trajectories.

As for the OUV model, the continuous time model version is given by:

$$\begin{aligned} & dr(t) = a(b - r(t))dt + \sigma\sqrt{r(t)}dB(t), \\ & r(0) = r_0. \end{aligned} \tag{7.44}$$

with the same assumptions as in the OUV model.

7.5.5.2. *Model and parameter interpretation*

We have here a nonlinear stochastic differential equation but for a null volatility, we still obtain the determinisitc OUV model.

7.5.5.3. *Conditional distribution of r(t), fixed t*

This nonlinearity implies that there does not exist a simple "explicit" form of the solution of the SDE [7.44]; nevertheless, the authors obtained the following explicit form of the conditional density function of $r(t)$, giving $r(s)$ $(s<t)$:

$$\begin{aligned} & f(y,t|r(s) = x) = Ke^{-K(u+v)}\left(\frac{v}{u}\right)^{q/2} I_q(2\sqrt{uv}), \\ & K = \frac{2a}{\sigma^2(1 - e^{-a(t-s)})}, \\ & u = Kxe^{-a(t-s)}, v = Ky, \\ & q = \frac{2ab}{\sigma^2} - 1. \end{aligned} \tag{7.45}$$

K_q being the modified Bessel function of second kind of order q already defined in Chapter 6.

From result [7.45], it is possible to obtain the following expressions for the conditional mean and variance of $r(t)$:

$$\begin{aligned} E\left[r(t)\middle|r(s)=x\right] &= xe^{-a(t-s)} + b(1-e^{-a(t-s)}), \\ \operatorname{var}\left[r(t)\middle|r(s)=x\right] &= x\frac{\sigma^2}{a}(e^{-a(t-s)} - e^{-2a(t-s)}) + b\frac{\sigma^2}{2a}(1-e^{-a(t-s)})^2. \end{aligned} \qquad [7.46]$$

REMARK 7.2.– These latter results immediately give the asymptotic forms of the mean and the variances as follows:

$$\begin{aligned} \lim_{t\to\infty} E\left[r(t)\middle|r(s)=x\right] &= b \\ \operatorname{var}\left[r(t)\middle|r(s)=x\right] &= b\frac{\sigma^2}{2a}. \end{aligned} \qquad [7.47]$$

These results show that the interpretation of the two parameters a and b is the same as for the OUV models. Let us remark that the conditional variance is inversely proportional to a.

Let us also point out that for $a \to 0$, we obtain for fixed t:

$$\begin{aligned} E\left[r(t)\middle|r(s)=x\right] &\to x, \\ \operatorname{var}\left[r(t)\middle|r(s)=x\right] &\to x\sigma^2(t-s). \end{aligned} \qquad [7.48]$$

REMARK 7.3.– The authors show that:

i) if $2ab \geq \sigma^2$, then the solution of the SDE [7.44] never becomes zero starting with a strictly positive value at time 0;

ii) if $2ab < \sigma^2$, then it is possible that the spot rate takes on a 0 value but it will never take negative values as it could be for the OUV model.

7.5.5.4. *Confidence interval for r(t), fixed t*

As the conditional distribution of $r(t)$ is given by result [7.45], it is possible to construct a confidence interval for this spot rate at time t.

Moreover, asymptotically, the authors proved that:

$$\lim_{t\to\infty} f(y,t|r(s)=x)=\frac{\theta^{\nu}}{\Gamma(\nu)}y^{\nu-1}e^{-\theta y},$$
$$\theta=\frac{2a}{\sigma^2},\nu=\frac{2ab}{\sigma^2}. \qquad [7.49]$$

Of course, this last result gives results [7.47] for the asymptotic mean and variances.

From the practical point of view, it is seen that the bounds of the confidence interval for fixed t quickly converge to the bounds of the asymptotic one.

EXAMPLE 7.11.–

r_0	0,04	b	0,02
A	5		
σ	0,07		
Δt	0,01		

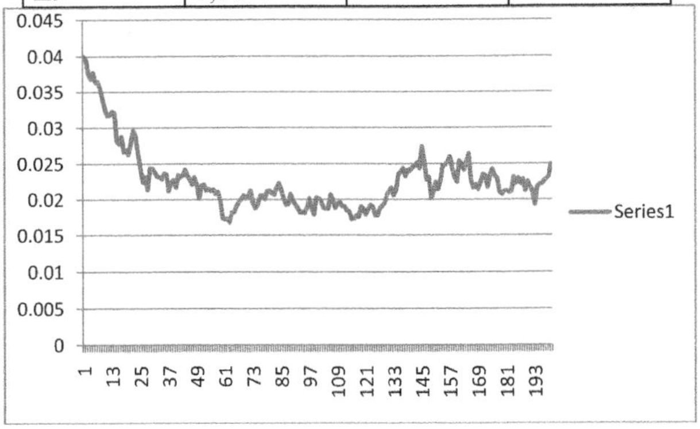

Figure 7.5. *A sample path for a CIR model*

7.6. Zero-coupon pricing under the assumption of no arbitrage

7.6.1. *Stochastic dynamics of zero-coupons*

As before, let $P(t,s)(t<s)$ represent at time t the value of a zero-coupon of time maturity s, thus, of maturity $T = t\text{-}s$ at time t. Let $T = s\text{-}t$ so that with our preceding notation:

$$P(t,s)=P(t,t+T) \qquad [7.50]$$

and so

$$P(s,s)=1. \tag{7.51}$$

The general problem of the evaluation of zero-coupons consists of studying the stochastic process P defined on the filtered probability space $(\Omega,\Im,(\Im_t),P)$ as follows:

$$P=\{P(t,s,\omega),t\in[0,s]\}. \tag{7.52}$$

Studying this process is equivalent to the study of process R of the equivalent instantaneous yields:

$$R=\{R(t,s,\omega),t\in[0,s]\}, \tag{7.53}$$

where:

$$\begin{aligned}P(t,s,\omega) &= e^{-(s-t)R(t,s-t,\omega)},\\ &= e^{-TR(t,T,\omega)}.\end{aligned} \tag{7.54}$$

Let us repeat that $R(t,T,\omega)$ is the *equivalent instantaneous rate* constant on the time interval $[t,t+T]$ given by present value, the value of the zero-coupon at time t, $P(t,T,\omega)$.

From this last relation, we can provide the value of R:

$$R(t,T,\omega)=-\frac{1}{T}\log P(t,t+T,\omega). \tag{7.55}$$

For the *spot* rate r, we have:

$$\lim_{T\to 0} R(t,T,\omega)=r(t,\omega). \tag{7.56}$$

The equivalent annual rate is given by:

$$e^{R(t,T,\omega)} = 1 + i(t,T,\omega). \qquad [7.57]$$

$$P(t,s,r_t) = \exp\left[-sR_\infty - \beta(s)(r_0 - R_\infty) - \frac{\sigma^2}{4a}\beta^2(s)\right],$$

where [7.58]

$$R_\infty = b + \frac{\lambda\sigma}{a} - \frac{\sigma^2}{2a^2},$$

$$r_0 = r(0).$$

Starting from a t different from 0, we obtain, from the Markov property:

$$P(t,s,r_0) = e^{-E_Q[U(t,s)] + \frac{1}{2}\mathrm{var}_Q[U(t,s)]}. \qquad [7.59]$$

and so:

$$P(t,s,r_t) = \exp\left[-(s-t)R_\infty - \beta(s-t)(r_t - R_\infty) - \frac{\sigma^2}{4a}\beta^2(s-t)\right],$$

where

$$\beta(u) = \frac{1-e^{-au}}{a}, \qquad [7.60]$$

$$R_\infty = b + \frac{\lambda\sigma}{a} - \frac{\sigma^2}{2a^2},$$

$$r_t = r(t).$$

Using result [7.55], we obtain the *instantaneous term structure*:

$$R(s,s-t) = R_\infty + \frac{\beta(s-t)}{s-t}(r_t - R_\infty) + \frac{\sigma^2}{4a}\frac{\beta^2(s-t)}{s-t}, s > t$$

$$\beta(u) = \frac{1-e^{-au}}{a}, \qquad [7.61]$$

$$R(t,\infty) = R_\infty,$$

$$E = R_\infty - \frac{\sigma^2}{4a^2}, F = R_\infty + \frac{\sigma^2}{2a^2}.$$

scen.	I	II	III	IV
par.	1-3,5%-3,5%-3%-1%	1-3,5%-3,5%-10%-1%	5-3,5%-3,5%-3%-1%	1-6%-3,5%-3%-1%
mat.	OUV	OUV	OUV	OUV
0.25	0.991281539	0.991281112	0.991282794	0.988627218
0.5	0.982633371	0.982690636	0.982637686	0.974895268
1	0.965571888	0.966061847	0.965571203	0.946419727
3	0.900418444	0.905690074	0.90021703	0.839356794
5	0.839775263	0.850918034	0.839286293	0.744380165
7	0.783233471	0.799721453	0.782479621	0.660150526
10	0.705481329	0.728695309	0.704396832	0.551334597
20	0.497890553	0.534459138	0.496171622	0.302451786
par.	a,b,r(0);sig;riskprm.			

Table 7.8. *Zero-coupons values with OUV*

EXAMPLE 7.12.–

Table 7.8 provides the result of zero-coupons values with OUV models with four scenarios given by the five parameters, selected as given by lines 2 and 3.

The most significant difference appears with the last column for which the retained scenario takes 6% for the long-term spot rate instead of 3,5% for the three other ones. In this case, the zero-bond with maturity 20 years has a value of 0.30 instead of more or less 0.50 for the first three scenarios.

7.6.2. *The CIR process as rate dynamic*

We can also use the CIR model presented in section 5.5.2:

$$\begin{aligned} dr(t) &= a(b-r(t))dt + \sigma\sqrt{r(t)}dB(t), \\ r(0) &= r_0. \end{aligned} \qquad [7.62]$$

We will express the premium risk in the form:

$$\lambda(t,r) = -\frac{\alpha}{\sigma}\sqrt{r(t)}. \qquad [7.63]$$

Cox *et al.* [COX 85] obtained the solution under the form:

$$P(r,t,s) = A(t,s)e^{-rD(t,s)},$$

where

$$A(t,s) = \left[\frac{2\gamma e^{\frac{(a+\alpha+\gamma)(s-t)}{2}}}{(a+\alpha+\gamma)(e^{\gamma(s-t)}-1)+2\gamma} \right]^{\frac{2ab}{\sigma^2}}, \quad [7.64]$$

$$D(t,s) = \frac{2(e^{\gamma(s-t)}-1)}{(a+\alpha+\gamma)(e^{\gamma(s-t)}-1)+2\gamma},$$

$$\gamma = \sqrt{(a+\alpha)^2 + 2\sigma^2}.$$

So, as

$$P(r,t,s) = e^{-(s-t)R(r,t,s)}, \quad [7.65]$$

we obtain from [7.64]:

$$R(r,t,s) = \frac{rD(t,s) - \ln A(t,s)}{(s-t)}. \quad [7.66]$$

The two limit cases give as results:

$$s \to t \Rightarrow R(r,t,s) \to r(=r(t)),$$
$$s \to \infty \Rightarrow R(r,t,s) \to \frac{2ab}{a+\alpha+\gamma}. \quad [7.67]$$

Table 7.9 provides the result of zero-coupon values with CIR models with four scenarios given by the five parameters, selected as given by lines 2 and 3.

scen.	I	II	III	IV
par.	1–3.5%–3.5%	1–3.5%–3.5%	5–3.5%–3.5%	1–6%–3.5%
	3%–1%	25%–1%	3%–1%	3%–1%
mat.	CIR I	CIR II	CIR III	CIR IV
0.25	0.99128664	0.99127952	0.99128701	0.98863114
0.5	0.98264681	0.98263459	0.98264889	0.97490507
1	0.96558803	0.96561683	0.96559715	0.94644076
3	0.90024357	0.90102057	0.90029773	0.8394155
5	0.83931463	0.84105891	0.83941429	0.74446771
7	0.78325086	0.78512914	0.78264807	0.66025964
10	0.70442635	0.7081375	0.70461487	0.55146515
20	0.49620034	0.5020042	0.49648109	0.30266638

Table 7.9. *Zero-coupon values with CIR*

Here too, as for the OUV example, the most significant difference appears with the last column for which the retained scenario takes 6% for the long-term spot rate instead of 3.5% for the three other ones. In this case, the zero-bond with maturity 20 years has a value of 0.30 instead of more or less 0.50 for the first three scenarios.

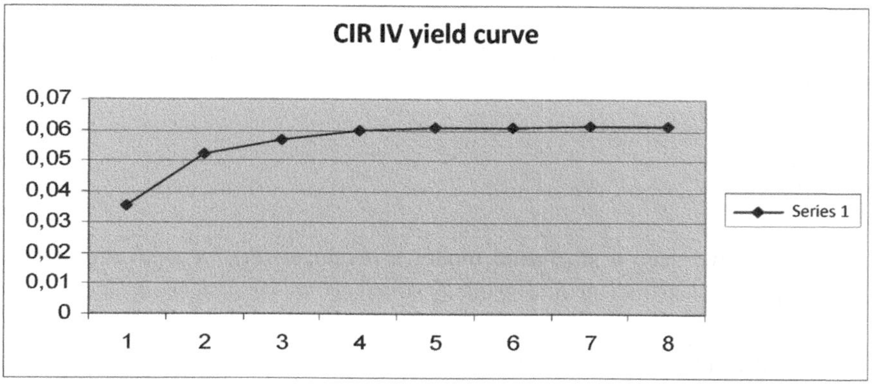

Figure 7.6. *Yield curve with the scenario CIR IV*

scen	I	II	III	IV
par.	1–3,5%–3,5%–3% –1%	1–3,5%–3,5%–10% –1%	5–3,5%–3,5%–3% –1%	1–6%–3,5%–3% –1%
mat.	OUV	OUV	OUV	OUV
0.25	0.991281539	0.991281112	0.991282794	0.988627218
0.5	0.982633371	0.982690636	0.982637686	0.974895268
1	0.965571888	0.966061847	0.965571203	0.946419727
3	0.900418444	0.905690074	0.90021703	0.839356794
5	0.839775263	0.850918034	0.839286293	0.744380165
7	0.783233471	0.799721453	0.782479621	0.660150526
10	0.705481329	0.728695309	0.704396832	0.551334597
20	0.497890553	0.534459138	0.496171622	0.302451786
par.	a,b,r(0);sig;riskprm.			

Table 7.10. *Spread OUV and CIR for the four scenarios*

Let us also point out that Table 7.10 shows that there is a good similarity between the numerical results of Tables 7.8 and 7.9 showing in these examples that, with similar scenarios, the OUV and CIR numerical results are closed, which is something of a *robustness result*.

7.7. Market evaluation of financial flows

As in section 7.1, let us consider the following financial flow:

$$((F_j, t_j), j = 1, \ldots, n)$$

To compute a future market value, for example at time t, we have to use the following formula:

$$\sum_{j=1}^{n} P(t,t_j)F_j \qquad [7.68]$$

Therefore, we need a stochastic model to evaluate $P(t,t_j)$ with the mark-to-model method.

The next example gives results using the four CIR scenarios for the yield curve with a flow with data given in Table 7.10.

Example	Flux	CIR	Present val.
mat.			
0.25	1000	scenario1	26561.718
0.5	1000	scenario2	26702.853
1	1000	scenario3	26565.274
3	3000	scenario4	21017.57
5	3000		
7	5000		
10	10000	max. spread	5685.283
20	15000	spread in %	0.27050144
		of min.	

Table 7.11. *Present values with the four CIR scenarios*

As a comparison, we also compute present values of the considered flow with two flat yield curves and Table 7.12 gives the result.

i=3%	28093.6250
i=6%	21247.1571

Table 7.12. *PV with flat yield curves*

Finally, we also consider the four OUV scenarios to evaluate present values for the same flow and we obtain the following comparative results.

Evaluation	OUV	CIR	Spread	in % OUV
scenario1	26599.4069	26561.7190	37.6878535	0.14168682
scenario2	27512.3053	26702.8530	809.452336	2.94214652
scenario3	26556.9424	26565.2745	-8.3320509	-0.0313743
scenario4	23898.1515	21017.5720	2880.57953	12.0535663

Table 7.13. *Evaluation comparison*

In conclusion, this example clearly shows that though the yield curve scenarios are not too different, we see that in the worst case, there is a difference of more or less 12%, a difference that, if a bad choice of scenario has been made, must be covered by the equities of the funds. This is why it is absolutely necessary to build hedging structure based on the concepts of *VaR* as imposed now as well by Basel II or III as Solvency II.

Moreover, the stochastic models presented in this chapter can be used in what we now call a *generator of economic scenarios* used particularly in stress tests to see if the equities of the insurance can support the impact of these scenarios particularly of course in the case of catastrophic events or at least of the occurrence of rare events giving strong financial impacts.

Let us also recall, as we have seen in Chapter 6, that there exist extensions of the two main models presented here when the Brownian parts in SDE [7.35] and [7.44] are replaced by a Lévy process. In this case, as we have shown in Chapter 6, we lose closed formulas and the best that can be done is to obtain the Fourier transform of price formulas and therefore we have to use the inverse Fourier transform together with numerical methods to obtain numerical values (see [ZHU 10]). Of course, simulation methods are still possible.

Bibliography

[AND 67] ANDERSEN E.S., "An algebraic treatment of fluctuations of sums of random variables", *Proc. 5th Berkeley Sympos. Math. Statist. and Probability* (Berkeley, California, 1965/66), University of California Press, Berkeley, vol. 2, pp. 423–429, 1967.

[APP 04] APPLEBAUM D., *Lévy Processes and Stochastic Calculus*, Cambridge University Press, 2004.

[ASM 00] ASMUSSEN S., *Ruin Probabilities*, Advanced Series on Statistical Science & Applied Probability, World Scientific Publishing Co., Inc., River Edge, NJ, vol. 2, 2000.

[BAK 77] BAKER C.T.H., *The Numerical Treatment of Integral Equations*, Clarendon Press, New York, 1977.

[BAR 62] BARLOW R.E., "Applications of semi-Markov processes to counter problems", *Studies in Applied Probability and Management Science*, Stanford University Press, Stanford, California, pp. 34–62, 1962.

[BAR 77] BARNDORFF-NIELSen O., HALGREEN C., "Infinite divisibility of the hyperbolic and generalized inverse Gaussian distributions", *Wahrscheinlichkeitstheorie und Verw. Gebiete*, vol. 38, no. 4, pp. 309–311, 1977

[BAR 77] BARNDORFF-NIELSEN O., "On conditional inference for deviation from Hardy-Weinberg distribution", in CHRISTIANSEN F.B., FENCHEL T. (eds), *Lecture Notes in Biomathematics*, Springer, vol. 19, pp. 149–157, 1977.

[BER 99] BERTOIN J., "Renewal theory for embedded regenerative sets", *Annals of Probability*, vol. 27, pp. 1523–1535, 1999.

[BLA 73] BLACK F., SCHOLES M., "The pricing of options and corporate liabilities", *Journal of Political Economy*, vol. 81, pp. 637–659, 1973.

[BRE 68] BREIMAN L., *Probability Theory*, Addison-Wesley, Reading, MA, 1968.

[CAR 99] CARR P., MADAN D.B., "Option valuation using the fast Fourier transform", *Journal of Computational Finance*, vol. 2, pp. 61–73, 1999.

[CHU 60] CHUNG K.L., *Markov Chain with Stationary Transition Probabilities*, Springer, 1960.

[CHU 00] CHUNG K.L., *A Course in Probability Theory*, 3rd ed., Academic Press, Inc., San Diego, 2000.

[ÇIN 69] ÇINLAR E., "Markov renewal theory", *Advances in Applied Probability*, vol. 1, pp. 123–187, 1969.

[ÇIN 75] ÇINLAR E., "Markov renewal theory: a survey", *Management Science*, vol. 21, pp. 727–752, 1975.

[CON 04] CONT R., TANKOV P., *Financial Modelling with Jump Processes*, Boca Raton, 2004.

[COR 04] CORRADI G., JANSSEN J., MANCA R., "Numerical treatment of homogeneous semi-Markov processes in transient case", *Methodology and Computing in Applied Probability*, vol. 6, pp. 233–246, 2004.

[COR 12] CORLOSQUET-HABART M., JANSSEN J., MANCA R., *Modélisation stochastique du risque de pandémie*, Hermes-Lavoisier, Paris, 2012.

[COX 65] COX D.R., MILLER H.D., *The Theory of Stochastic Processes*, John Wiley & Sons, Inc., New York, 1965.

[COX 85] COX J.C., INGERSOLL J.E., ROSS S.A., "A theory of the term structure of interest rates", *Econometrica*, vol. 53, pp. 385–407, 1985.

[CRA 55] CRAMÉR H., *Collective Risk Theory*, Skandia Jubilee volume, Stockholm, 1955.

[DAH 72] DAHL O.J., DIJKSTRA E.W., HOARE C.A.R., *Structured programming*, Academic Press, London, 1972.

[DAM 08] D'AMICO G., BIFFI E., DI BIASE G. *et al.*, "Monte Carlo semi-Markov methods for credit risk migration models and Basel II rules I & II part", *Journal of Numerical and Applied Mathematics*, vol. 96, pp. 28–58 & 59–86, 2008.

[DAM 09a] D'AMICO G., JANSSEN J., MANCA R., "The dynamic behavior of single unireducible non-homogeneous Markov and semi-Markov chains", in NAIMZADA A., STEFANI S., TORRIERO A. (eds), *Networks: Topology and Dynamic*, Lectures Notes in Economic and Mathematical Systems, Springer, pp. 195–211, 2009.

[DAM 09b] D'AMICO G., JANSSEN J., MANCA R., "European and American options: the semi-Markov case", *Physica A*, vol. 388, pp. 3181–3194, 2009.

[DED 84] DE DOMINICIS R., MANCA R., "A computational procedure for the asymptotic analysis of a homogeneous semi Markov process", *Statistics & Probability Letters*, vol. 2, pp. 249–253, 1984.

[DEV 96] DE VYLDER F., MARCEAU E., "Classical numerical ruin probabilities", *Scandinavian Actuarial Journal*, pp. 109–123, 1996.

[DEV 09] DEVINEAU L., LOISEL S., "Risk aggregation in Solvency II: how to converge the approaches of the internal models and those of the standard formula?", *Bulletin Français d'Actuariat*, vol. 18, pp. 107–145, 2009.

[DEV 12] DEVOLDER P., JANSSEN J., MANCA R., *Stochastic Methods for Pension Funds*, ISTE Ltd, London and John Wiley & Sons, New York, 2012.

[DIB 97] DI BIASE G., JANSSEN J., MANCA R., "Markov and semi-Markov option pricing models with arbitrage possibility", *Applied Stochastic Models and Data Analysis*, vol. 13, pp. 103–113, 1997.

[DOO 48] DOOB J.L., "Renewal theory from the point of view of the theory of probability", *Transaction of American Mathematical Society*, vol. 63, pp. 422–438, 1948.

[DRO 08] DROGUETT E.L., MOURA M.C., JACINTO C.M.C., "A semi-Markov model with Bayesian belief network based human error probability for availability assessment of downhole optical monitoring systems", *Simulation Modeling Practice Theory*, vol. 16, pp. 1713–1727, 2008.

[DRO 09] DROGUETT E.L., MOURA M.C., "Mathematical formulation and numerical treatment based on transition frequency densities and quadrature methods for non-homogeneous semi-Markov processes", *Reliability Engineering & System Safety*, vol. 94, pp. 342–349, 2009.

[EBE 95] EBERLEIN E., KELLER U., "Hyperbolic distributions in finance", *Bernoulli*, vol. 1, pp. 281–299, 1995.

[EBE 98] EBERLEIN E., KELLER U., PRAUSE K., "New insights into smile, mispricing, and value at risk: the hyperbolic model", *The Journal of Business*, vol. 71, pp. 371–405, 1998.

[EBE 99] EBERLEIN E., RAIBLE S., "Term structure models driven by general Lévy processes", *Mathematical Finance*, vol. 9, pp. 31–53, 1999.

[ELK 01] ELKINS D.A., WORTMAN M.A., "On numerical solution of the Markov renewal equation: tight upper and lower kernel bounds", *Methodology and Computing in Applied Probability*, vol. 3, pp. 239–253, 2001.

[ELK 11] EL KAROUI N., GOBET E., *Les outils stochastiques des marchés financiers, Une visite guidée de Einstein à Black-Scholes*, Les Éditions de l'École polytechnique, 2011.

[EMB 08] EMBRECHTS P., KLÜPPELBERG C., MIKOSCH T., *Modelling Extremal Events for Insurance and Finance*, Springer, 2008.

[ESC 97] ESCH L., KIEFFERR., LOPEZ T., *Value at Risk*, De Boeck Université, Brussels, 1997.

[FEL 41] FELLER W., "On the integral equation of renewal theory", *Annals of Mathematical Statistics*, vol. 12, pp. 243–267, 1941.

[FEL 57] FELLER W., *An Introduction to Probability Theory and its Applications*, vol. 1, 2nd ed., Wiley, New York, pp. 461, 1957.

[FEL 68] FELLER W., *An Introduction to Probability Theory and its Applications*, 1st ed., John Wiley & Sons, Inc., New York-London-Sydney, vol. 2, 1968.

[FEL 71] FELLER W., *An Introduction to Probability Theory and its Applications*, 2nd ed., John Wiley & Sons, Inc., New York-London-Sydney, vol. 2, 1971.

[FIS 30] FISHER I., *The Theory of Interest*, MacMillan Co, New York, 1930.

[FRE 71] FREIBERGER W., GRENANDER U., *A Short Course in Computational Probability and Statistics*, Springer Verlag, Berlin, vol. 12, p. 155, 1971.

[FRI 75] FRIEDMAN A., *Stochastic Differential Equations and Applications*, 2nd ed., Academic Press, New York, 1975.

[GAR 88] GARD C., *Introduction to Stochastic Differential Equations*, Marcel Dekker, 1988.

[GEH 15] GEHIN W., CORLOSQUET-HABART M., JANSSEN J. *et al.*, *Asset and Liabilities Management for Banks and Insurance Companies,* ISTE Ltd., London and John Wiley & Sons, New York, 2015.

[GER 97] GERBER H.U., SHIU ELIAS S.W., "The joint distribution of the time of ruin, the surplus immediately before ruin, and the deficit at ruin", *Insurance Mathematics and Econonomics*, vol. 21, pp. 129–137, 1997.

[GIK 69] GIKHMAN I.I., SKOROHOD A.V., *Introduction to the Theory of Random Processes I*, Translated from the Russian by HIRSCH W.M., SAUNDERS W.B., Philadelphia, 1969.

[GIK 75] GIHMAN I.I., SKOROHOD A.V., *The Theory of Stochastic Processes I,* Translated from the Russian by KOTZ S., Springer Verlag, Berlin, New York, 1975.

[GIS 15] GISMONDI F., JANSSEN J., MANCA R., "Non-homogeneous time convolutions, renewal processes and age-dependent mean number of motor car accidents", *Annals of Actuarial Sciences*, vol. 9, pp. 36–57, 2015.

[GNE 43] GNEDENKO B.V., "On the growth of homogeneous random processes with independent single-type increments", *C.R. Doklady Academy Science URSS*, vol. 40, pp. 90–93, 1943.

[HAL 74] HALMOS P.R., *Measure Theory*, Springer, New York, 1974.

[HIL 87] HILDEBRAND F.B., *Introduction to Numerical Analysis*, 2nd ed., Dover, New York, 1987.

[HOL 96] HOLDEN H., ØKSENDAL B., UBØE J. *et al.*, *Stochastic Partial Differential Equations – a Modeling, White Noise Functional Approach*, Birkhauser, Boston, 1996.

[HOW 71] HOWARD R., *Dynamic Probabilistic Systems*, 2nd ed., Wiley, New York, 1971.

[ISA 76] ISAACSON D.L., MADSEN R.W., *Markov Chains, Theory and Applications*, Wiley Series in Probability and Mathematical Statistics, John Wiley & Sons, New York-London-Sydney, 1976.

[ITÔ 44] ITÔ K., "Stochastic integral", *Proc. Imp. Acad.*, Tokyo, vol. 20, pp. 519–524, 1944.

[JAN 69] JANSSEN J., "Les Processus(J-X)", *Cahiers du C.E.R.O.*, vol. 11, pp. 181–214, 1969.

[JAN 70] JANSSEN J., "Sur une Généralisation du Concept de Promenade Aléatoire sur la Droite Réelle", *Annales de l' Institut H. Poincaré*, vol. 6, pp. 249–269, 1970.

[JAN 77] JANSSEN J., "The semi-Markov model in risk theory", in ROUBENS M. (ed.), *Advances in Operations Research*, North-Holland, Amsterdam, pp. 613–621, 1977.

[JAN 86] JANSSEN J. (ed.), *Semi-Markov Models. Theory and Applications*, Plenum Press, New York, 1986.

[JAN 99a] JANSSEN J., LIMNIOS N. (eds), *Semi-Markov Models and Applications*, Kluwer Academic, New York, 1999.

[JAN 99b] JANSSEN J., MANCA R., VOLPE DI PRIGNANO E., "Option pricing with semi-markov volatility", in JANSSEN J., LIMNIOS N. (eds), *Semi-Markov Models and Applications*, Kluwer Academic, New York, 1999.

[JAN 01] JANSSEN J., MANCA R., "Numerical solution of non homogeneous semi-Markov processes in transient case", *Methodology and Computing in Applied Probability*, Kluwer, vol. 3, pp. 271–293, 2001.

[JAN 06] JANSSEN J., MANCA R., *Applied Semi-Markov Processes*, Springer, New York, 2006.

[JAN 07] JANSSEN J., MANCA R., *Semi-Markov Risk Models for Finance, Insurance and Reliability*, Springer Verlag, New York, 2007.

[JAN 09a] JANSSEN J., MANCA R., *Outils de construction de modèles internes pour les assurances et les banques*, Hermes-Lavoisier, Paris, 2009.

[JAN 09b] JANSSEN J., VOLPE DI PRIGNANO E., MANCA R., *Mathematical Finance. Deterministic and Stochastic Models*, ISTE Ltd, London and John Wiley & Sons, New York, 2009.

[KAR 81] KARLIN S., TAYLOR H.M., *A Second Course in Stochastic Processes*, Academic Press, New York, 1981.

[LIM 01] LIMNIOS N., OPRIŞAN G., *Semi-Markov Processes and Reliability*, Birkhauser, Boston, 2001.

[LIM 03] LIMNIOS N., OPRIŞAN G., "An introduction to semi-Markov processes with application to reliability", in SHANBHAG D.N., RAO C.R. (eds), *Handbook of Statistics*, vol. 21, pp. 515–556, North-Holland, Amsterdam, 2003.

[LOÈ 63] LOÈVE M., *Probability Theory*, 3rd ed., Princeton, New York, 1963.

[LOÈ 77] LOÈVE M., *Probability Theory*, 4th ed., Springer, New York, 1977.

[LOT 74] LOTKA A.J., "Sur une équation intégrale de l'analyse démographique et industrielle", *Mitt. Verein. Schweiz. Versich.-Math.*, vol. 40, pp. 1–16, 1940.

[LUK 70] LUKACS E., *Characteristic Functions*, Griffin's Statistical Monographs & Courses, no. 14, Hafner Pub. Co., New York, 1970.

[LUN 09] LUNDBERG P., "On the theory of reinsurance", in HABERMANN S., SIBBETT T.A. (eds), *History of Actuarial Science*, 7th ed., 1995, pp. 291–244, 1909.

[MAD 90] MADAN D.B., SENETA E., "The variance gamma (V.G.) model for share", *Journal of Computational Finance*, vol. 2, pp. 61–73, 1990.

[MAD 98] MADAN D.B., CARR P., CHANG E.C., "The variance gamma process and option pricing", *European Finance Review*, vol. 2, no. 79, p. 105, 1998.

[MAN 62] MANDELBROT B.B., "Sur certains prix spéculatifs: faits empiriques et modèle basé sur les processus stables additifs de Paul Lévy", *Comptes rendus* (*Paris*), vol. 254, pp. 3968–3970, 1962.

[MAO 97] MAO X., *Stochastic Differential Equations and Applications*, Horwood, New York, 1997.

[MAR 59] MARKOWITZ H.M., *Portfolio Selection: Efficient Diversification of Investments*, Cowles Foundation for Research in Economics at Yale University, Monograph 16, John Wiley & Sons, Inc., New York, 1959.

[MER 73] MERTON R.C., "An inter temporal capital asset pricing model", *Econometrica*, vol. 41, pp. 867–887, 1973.

[MER 76] MERTON R.C., "Option pricing when underlying stock returns are discontinuous", *Journal of Financial Economics*, vol. 3, pp. 125–144, 1976.

[MEY 66] MEYER P.A., *Probabilités et Potentiel*, Hermann, Paris, 1966.

[MIK 09] MIKOSCH T., *Non-Life Insurance Mathematics*, Springer, Berlin, 2009.

[MIL 62] MILLER H.D., "Absorption probabilities for sums of random variables defined on a finite Markov chain", *Proceedings of Cambridge Philosophic Society*, vol. 58, pp. 286–298, 1962.

[MIY 12] MIYAHARA Y., *Option Pricing in Incomplete Markets: Modeling Based on Geometric Levy Processes and Minimal Entropy Martingale Measures*, Imperial College Press, London, 2012.

[MOU 10] MOURA M.C., DROGUETT E.L., "Numerical approach for assessing system dynamic availability via continuous time homogeneous semi-Markov processes", *Methodology and Computing in Applied Probability*, vol. 12, pp. 431–449, 2010.

[OSA 85] OSAKI S., *Stochastic System Reliability Modelling*, Series in Modern Applied Mathematics, World Scientific Publishing Co., Singapore, vol. 5, 1985.

[PAR 62] PARZEN E., *Stochastic Processes*, Holden-Day Series in Probability and Statistics, San Francisco, 1962.

[PAS 11] PASCUCCI A., *PDE and Martingale Methods in Option Pricing*, Springer, 2011.

[PLA 05] PLANCHET F., THEROND P.E., JUILLARD M., "Modèles financiers en assurance. Analyses de risques dynamiques", 2nd ed., *Economica*, Paris, 2005.

[PLA 06] PLATEN E., HEATH D., *A Benchmark Approach to Quantitative Finance*, Springer, New York, 2006.

[PRO 90] PROTTER P.E., *Stochastic Integration and Differential Equations*, A New Approach, Applications of Mathematics, Springer Verlag, Berlin, 1990.

[PYK 62] PYKE R., "Markov renewal processes of zero order and their application to counter theory", *Studies in Applied Probability and Management Science*, Stanford University Press, Stanford, California, pp. 173–183, 1962.

[RIE 13] RIESZ F., *Les systèmes d'équations linéaires à une infinité d'inconnues*, Gauthiers-Villars, Paris, 1913.

[SAM 65] SAMUELSON P.A., "Rational theory of warrant pricing", *Industrial Management Review*, vol. 6, pp. 13–31, 1965.

[SAT 99] SATO K., *Lévy Processes and Infinitely Divisible Distributions*, Cambridge Studies in Advanced Mathematics, Cambridge University Press, vol. 68, 1999.

[SCH 03] SCHOUTENS W., *Levy Processes in Finance: Pricing Financial Derivatives*, Wiley, 2003.

[SMI 54] SMITH W.L., "A symptotic renewal theorems", *Proceedings of the Royal Society of Edinburgh*, vol. 64, pp. 9–48, 1954.

[SMI 58] SMITH W.L., "Renewal theory and its generalizations", *Journal of the Royal Statistical Society B*, vol. 20, pp. 243–302. 1958.

[XIE 89] XIE M., "On the solution of renewal-type integral equations", *Communications in Statististics–Simulation and computation*, vol. 18, pp. 281–293. 1989.

[ZHU 00] ZHU J., *Modular Pricing Options: an Application of Fourier Analysis*, Springer, Berlin, 2000.

[ZHU 10] ZHU J., *Applications of the Fourier transform to Smile Modeling*, 2nd edition, Springer Verlag, Heidelberg, 2010.

Index